D1212726

Chrambach
The Practice of
Quantitative
Gel Electrophoresis

VCH

Advanced Methods in the Biological Sciences

Edited by V. Neuhoff and A. Maelicke

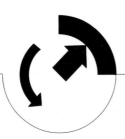

Distribution:

VCH Verlagsgesellschaft, P.O. Box 1260/1280, D-6940 Weinheim (Federal Republic of Germany)

USA and Canada: VCH Publishers, 303 N.W. 12th Avenue, Deerfield Beach, FL 33442-1705 (USA)

ISBN 3-527-26039-0 (VCH Verlagsgesellschaft)
ISBN 0-89573-064-2 (VCH Publishers)

ISSN 0178-9244

Andreas Chrambach

The Practice of Quantitative Gel Electrophoresis

VCH

Dr. A. Chrambach
Department of Health
and Human Services
National Institutes of Health
Bethesda MD 20205, USA

Prof. Dr. V. Neuhoff
Max-Planck-Institut für
experimentelle Medizin
Hermann-Rein-Str. 3
D-3400 Göttingen

Prof. Dr. A. Maelicke
Max-Planck-Institut für
Ernährungsphysiologie
Rheinlanddamm 201
D-4600 Dortmund

1st edition 1985

Editorial Director: Dr. Hans F. Ebel

Deutsche Bibliothek Cataloguing-in-Publication-Data

Chrambach, Andreas:
The practice of quantitative gel electrophoresis / Andreas Chrambach. – Weinheim; Deerfield
Beach, FL: VCH, 1985.
  (Advanced methods in the biological sciences)
  ISBN 3-527-26039-0 (Weinheim)
  ISBN 0-89573-064-2 (Deerfield Beach, FL)
  ISSN 0178-9244
NE: GT

Composition: kühn & weyh software GmbH, D-7800 Freiburg
Printing: Zechnersche Buchdruckerei, D-6720 Speyer
Bookbinding: Klambt-Druck GmbH, D-6720 Speyer
Cover Design: Klaus Grögerchen, D-6901 Lampenhain
Printed in the Federal Republic of Germany

Preface

The topic of the book is gel electrophoresis as a *unity*. This topic comprises polyacrylamide and agarose gel electrophoresis, isotachophoresis and electrofocusing. Our attempt will be to show that all of these are not competing methods of separation, but rather that each of them has its well defined area of applicability and provides the greatest efficiency of separation only within this unique area.

The electrophoretic separations for which the methods described in this book are preponderantly designed are those of *native* macromolecules, in contradistinction to the dissociated and denatured gene products. This bias derives from the axiom that biochemistry aims at an account of biological function in terms of chemical structures and reactions. Necessarily, these structures comprise the sum of post-translational molecular forms which participate in a given enzymatic, receptor, hormonal or other function.

A presentation of the quantitative, *theory-based* approaches to the separation of charged molecular species by gel electrophoresis is useful because these rational approaches provide a greater efficiency of separation than the arbitrarily selected methods. However, the sure way to prevent the average biochemist from using these (as any other) theory-based approaches in practice is to present the theoretical treatments. Therefore, this book contains no separation theory. It is exclusively a *practical* guide to separation of proteins or other charged molecular species. The results of theoretical considerations, leading to optimal efficiency of separation, are being provided only in form of computer output and the simple manipulations needed to obtain it. It is, in fact, this unique feature of computer programs of being able to imbed complex theory while providing results simply, that has made it possible for the average biochemist – most of us – to reap the fruit of separation theory in providing a more efficient separation practice.

A philosophical point needs to be made concerning this practical, non-theoretical presentation. It has to be realized that in areas such as the regulation of the concentrations, pH and other physical parameters of the trailing phase by the leading phase in moving boundary electrophoresis, a real understanding can only be conveyed by theoretical and mathematical language, i.e. in a commonly un-understandable form. To the degree that a qualitative, approximative and descriptive approach is used, the full truth is not being presented. But – the presentation becomes also commonly understandable. Since communication is needed to serve the average biochemist in his separation problems, this book has in those instances deliberately sacrificed scientific rigor and depth of presentation. It risks, therefore, to be considered inadequate by the theoretician. I hope, nonetheless, that it will provide approximative useful insights to the practitioner in search of more effective separation methods.

The separation methods of interest to this book are exclusively *general* ones, which are equally applicable to the physical identification and isolation of any charged macromolecule. This restriction is due to the fact that in most applications, there is no information prior to separation concerning a specific property of the macromolecule by which it would distinguish itself from the majority of other species. To find such a property would require a trial-and-error approach, or playing for luck, which is precisely the experimental approach which the rational, systematic and quantitative methods outlined in this book attempt to overcome. Thus, specific affinity methods will not be treated, although clearly these, and any other separation method exploiting a specific peculiarity or composition of the macromolecule of interest, are preferable to general experimental approaches if and when such a specific handle for fractionation is known.

Although this book is exclusively a practical guide to separation, it should not be read linearly from beginning to end as one would read a recipe. The reason for this is that our particular recipe deals with terminology, with polymerization chemistry, with particular forms of apparatus, the detailed description and discussion of which necessarily interrupts the narrative. The reader is therefore advised to use this book as a *handbook*, by selecting from the Table of Contents those sections dealing with the separation steps which he wants to apply to his problem. Then, facing terminology, polymer chemistry or apparatus he is unfamiliar with, he should trace back by means of the Table of Contents to the particular pages dealing with those points to obtain the necessary information.

I should explain why the *apparatus* needed to do quantitative gel electrophoresis is described and critically discussed in exhaustive detail. This would not be necessary if at this time such apparatus were commercially available. However, only some items of the needed apparatus are available, and nearly always in imperfect form. Other items need to be constructed or assembled by the user. The detailed critical review of the equipment thus serves as a guide for those intent on building or improving their own apparatus. At the same time, it serves, hopefully, as a reminder to the manufacturers of instrumentation that the equipment needs of efficient gel electrophoresis are still largely unfilled, or remain filled in non-optimal fashion.

This book does not attempt an *exhaustive* presentation of separation by gel electrophoretic methods. Rather, an attempt is being made to provide the reader with the same *sufficient set* of separation tools (rationales, apparatus, procedures, computer programs) that have made it possible to efficiently characterize and isolate diverse proteins in the author's laboratory. Necessarily, this approach is limited by the knowledge and insight of the author, but it is *practical* in the sense that the tools described here have been tried out and work.

The credits given to authors and instrument manufacturers necessarily omit those whose work I have been unfamiliar with, and whose ideas and findings should have found their way into this book but didn't. To remedy the situation in future editions of this or similar works, let me ask you, the reader, to communicate any such omissions and rectifications to me.

Bethesda,
in June 1984

A. Chrambach

Contents

Appendixes   211

Glossary of Terms

ALPHA	upper buffer phase
BETA	stacking-phase buffer, as prepared
Bis	N,N'-methylenebisacrylamide
BV	buffer value
C1	concentration of constituent 1 (M)
C2	concentration of constituent 2 (M)
C3	concentration of constituent 3 (M)
C6	concentration of constituent 6 (M)
%C	crosslinking agent (g/100 mL) $\times$ 100/%T
CMC	critical micelle concentration
CZE	continuous zone electrophoresis
DATD	N,N'-diallyltartardiamide
EDA	ethylenediacrylate
EF	electrofocusing
Ferguson plot	plot of log (R_f) vs. %T
GAMMA	resolving-phase buffer, as prepared
i	current (amps)
I	ionic strength
IEF	isoelectric focusing
ITP	isotachophoresis
K_R	retardation coefficient [$-$ d log (R_f)/d%T]
κ	specific conductance $(1/\text{ohm} \cdot \text{cm}) \cdot 10^{-6}$
KP	potassium persulfate
LAMBDA	buffer phase containing constituent 2 displacing constituent 3
M_o	free electrophoretic mobility (cm^2/s/volt)
MBE	moving boundary electrophoresis
MW	molecular weight
NU, v	boundary displacement
PAGE	polyacrylamide gel electrophoresis
PI	operative resolving phase containing constituent 1 (set with constituent 3)
pI	isoelectric point
PCA	perchloric acid
r	resistivity (ohm cm)
R	ionic mobility relative to Na$^+$ (in computer output)
$\overline{R}$	geometric mean radius (nm)
R_f	electrophoretic mobility relative to front moving boundary
RM	net mobility relative to Na$^+$

RM(1,ZETA)   RM of constituent 1 in phase ZETA, trailing ion net mobility, stacking phase

RM(2,BETA)   RM of constituent 2 in phase BETA, leading ion net mobility, trailing phase

RM(1,PI)   RM of constituent 1 in phase PI, trailing ion net mobility, resolving phase

RM(1,4)   computer output designation for RM(1,ZETA)

RM(2,2)   computer output designation for RM(2,BETA)

RM(1,9)   computer output designation for RM(1,PI)

RN   riboflavin

SCAM   synthetic carrier ampholyte mixture

SDS   sodium dodecylsulfate

%T   total gel concentration (acrylamide plus Bis) (g/100 mL)

TCA   trichloroacetic acid

TEMED N,N,N',N'-tetramethylethylenediamine

Y_o extrapolated R_f at %T = 0 on the Ferguson plot

1 The Concept of an Objectively Defined Optimal Fractionation Route

This book deals with the separation between charged molecules in an "objectively defined" way which is distinct for each separation problem. It will be shown that the data are capable of dictating the method in each particular application. Furthermore, the method objectively defined for a particular separation will be shown to be optimal, i.e. to provide the greatest possible degree of efficiency. Thus, the mode of selecting the separation tool advocated here contrasts with the arbitrary choice of separation conditions common in biochemistry today.

The methods we will be dealing with all comprise one or the other form of *gel electrophoresis*. Although such limitation to gel electrophoretic methods appears arbitrary on its surface, in violation of the "objectivity" of methods definition we have just talked about, it is justified for our particular time in history. At this time (and the situation may already be quite the opposite tomorrow), gel electrophoresis has the advantage over nearly all other separation tools in being responsive, at the same time, to differences among molecular species with regard to *all* of their universal physical properties simultaneously, i.e. *molecular net charge, size* and *relative hydrophobicity*. Any attempt at optimizing the separation conditions must exploit all of these differences between species, either successively or simultaneously. Although common practice in biochemistry prefers to set up successive fractionation stages for separation on the basis of either one of the 3 classes of molecular properties, gel electrophoresis allows one to do the same thing in a single step, or at least by use of a single tool, within certain limits with regard to the complexity of the macromolecular system (see Section 10), and to load. Within these limits, gel electrophoresis is a sufficient tool, both analytically and preparatively. This notion of method *sufficiency* obviously contrasts with the traditional notion in biochemistry which postulates that the largest possible number of diverse fractionation tools should be applied to adequately separate macromolecules on the basis of these three properties.

It may be argued that DEAE-agarose, or DEAE-Sephadex, just like gel electrophoresis allow one to separate on the basis of molecular size and net charge differences simultaneously. Qualitatively, this is the case indeed. The difference between gel electrophoresis and those chromatographic methods relates to the degree to which the size of the molecular sieve responsible for discrimination on the basis of molecular size and shape can be adapted to the need of a particular separation and can thus be optimized.

Methods *optimization* is another point of departure between the approaches to separation outlined in this book and the conventional approaches. Under the optimization of a separation method, we understand both the selection of the most efficient separation tool, as well as the application of that tool with the greatest possible efficiency. Thus, for instance, many problems are solved more efficiently by gel electrophoresis than by gel electrofocusing or isotachophoresis, or the reverse. Furthermore, within gel electrophoresis, separation is most efficient at a particular pH and a particular gel concentration. How are these optimal conditions, and the optimal method, found? The answer is: By use of computer programs which incorporate, and make available to the average biochemist, the results of theoretical separation science and of statistics. To contrast the methods outlined in this book, which attempt by computations based on theory to objectively define the most efficient method, with the conventional use of gel electrophoresis we have coined for them the term "Quantitative" gel electrophoresis. Since at this time polyacrylamide is the gel matrix with the widest range of continuously variable pore sizes which are effective for separating most macromolecules, the term "Quantitative Polyacrylamide Gel Electrophoresis", abbreviated as "Quantitative PAGE" is also used. In application to viruses and large DNA species, a "Quantitative Agarose Electrophoresis" has also more recently been developed (Chapter 6).

Parenthetically, a philosophical point in regard to the use of computers should be made. By and large, biochemists still shrink from the use of computer programs in their daily work for two reasons, both erroneous: The first is the Renaissance notion that a scientist must completely understand everything he is doing. That includes computer programs as well as the theoretical and statistical body of knowledge upon which these programs rest. In our view, this classical notion negates precisely the importance of computers in contemporary science with its large numbers of investigators of necessarily mediocre caliber (the author included). The unique importance of computers in science is precisely that it is *not* necessary to know either the programs or their underlying theory to arrive at accurate results which without use of computers would have been available to only a very few theoreticians and statisticians. This is quite analogous to the use of complex instrumentation in the laboratory, of which we ordinarily also don't understand more than possibly a few of the underlying physical principles.

The second reason for a still widespread resilience to the use of computers in the laboratory is the erroneous notion that their use is complex, because computers or programs are complex. In fact the use of computers is so simple that it can be learned by most grammar school children, and certainly almost anyone working in the laboratory, without the slightest knowledge of how computers are constructed or programs written. Nonetheless, any intellectual realization of this fact does not seem sufficiently convincing; what it takes is to sit down by the terminal at the end of a workday, having just run a gel electrophoretic separation, stained the gels (in 1.5 h), measured the migration distances of the protein of interest at several gel concentrations, and to experience just once the wealth of physical and statistical information obtainable by computer from a morning's work. This cathartic experience will necessarily become common once computer time sharing terminals or minicomputers are stationed in the laboratory, as they will undoubtedly be universally within the next few years.

2 Selection of a Solvent for Gel Electrophoresis

Gel electrophoresis is mostly applied to the separation of *water soluble* species. It is usually applied in aqueous buffers at pHs varying from 3 to 11, and at ionic strengths varying from 0.01 to 0.03 M. It can equally well be applied outside of those ranges of pH and ionic strength if one can compensate for the high conductivity of buffer at an extreme of pH and at high ionic strength, and thus for the proportional amounts of Joule heat, by an increase in voltage and corresponding improvement of the heat dissipation capacity of the apparatus (Chapter 4). If this is not possible one can still carry out gel electrophoresis at conventional voltages and with apparatus of ordinary heat dissipation capacity if time dependent diffusion does not prevent one from conducting electrophoresis very slowly, possibly over a number of days. This is the case if the protein is large and asymmetric in shape, and/or the gel relatively confining. Since a restrictive gel further decreases the low migration rate in a shallow voltage gradient, a compromise between gel concentration and allowable time is required. This is practicable, as shown e.g. by the possibility of conducting gel electrophoresis of myosin in 0.6 M KCl for 1 week (oral communication of A. d'Albis in conjunction with Abstr.1d4, *9th Internatl. Congr. Biochem.*, Stockholm, 1973).

A second limitation exists with respect to the use of an extreme of pH in application to polyacrylamide gels, the amide bond in polyacrylamide being labile under those conditions. This does not prevent one, however, from applying at an extreme of pH agarose or other gel media at concentrations which yield pore sizes equivalent to those of polyacrylamide (Chapters 5,6).

Gel electrophoresis is, however, not limited to the separation of water soluble species. It is important to realize that the *water insoluble* macromolecules, or those present in associated states, are equally amenable to separation by gel electrophoretic fractionations. To stress this applicability to every kind of biological system, we will start by considering a problem which exists only in a minority of applications: The methods available for solubilizing water-insoluble species with maintenance of biological function. For application to water-insoluble species, a suitable solvent compatible with gel electrophoresis must be found. Furthermore, if native conformation, native molecular activity or any other native property is to be maintained, this solvent must be selected with care to preserve at least a substantial part of that function. In the water-insoluble form, a protein may be associated either to homologous species (aggregation) or heterologously, so as to require dissociation prior to gel electrophoresis. The bond responsible for the association may be hydrophobic, electrostatic or a H-bond, and thus the solvent should comprise dissociating agents specific for each of these bonds.

The choice of solubilizing medium should be made in a *systematic* fashion [1]–[3]. The first decision, in making that choice, is whether the bonds, by which the species of interest in its insoluble form is bound to either a particular matrix, or to other heterologous or homologous molecular species, is preponderantly of a *hydrophilic* or of a *hydrophobic* nature. The answer to that question is not necessarily related to the hydrophilic or hydrophobic character of the species, e.g. of a membrane protein, since hydrophobic proteins are capable of interactions across their hydrophilic domains, and *vice versa* hydrophilic proteins can interact across their hydrophobic domains (as in hydrophobic chromatography).

2.1   Systematic Incubation Experiments

2.1.1   Choice of Buffer

Experimentally, one would incubate the material in buffers of high, low and neutral pHs, containing 0, 0.1, and 0.5 M KCl to test for solubilization by dissociation of hydrophilic bonds alone. The preferred temperature for the maintenance of most protein functions is $0 - 4 °C$. The soluble and insoluble protein is separated either by centrifugation at 105000 *g*, or by chromatography on Sepharose 4-B, to name some popular methods used as criteria for solubility. If these media fail to solubilize, one may test the same series in presence of 0.1 M EDTA in order to ascertain that the insolubility is not due to bonding by divalent metal ions. Urea in concentrations up to 9 M may be required to sever hydrogen (and hydrophobic) bonds. Finally, polyols are frequently useful agents for solubilization and maintenance of macromolecular function, e.g. 25% glycerol. If all of these measures fail to solubilize, the additional disruption of hydrophobic bonds by detergents is indicated. It is predominantly through the admixture of detergents that the differences in relative hydrophobicity among macromolecules can be exploited for the purpose of separation. Only to a minor degree can this be done through use of relatively hydrophobic buffers, such as N-ethyl- or hydroxyethyl-morpholine.

2.1.2   Choice of Detergent

Relatively few types of detergent have proven to be effective solubilizing agents for water-insoluble proteins with maintenance of their functions: Cholate, CHAPS (Calbiochem No. 220201, Serva No. 17038), digitonin, sulfobetaine (e.g. Zwittergent 3-14, Calbiochem No. 693017), β-octylglucoside (Calbiochem No. 494459), Triton X-100 (alias NP-40, available in pure form as OPE-10, Calbiochem. No. 497020) and Lubrol PX. Since later electrophoretic fractionation is intended, the 6 uncharged non-ionic or amphoteric detergents in that series are to be preferred. Also preferred are those detergents with high ($<$ 1 mM) critical micelle concentration (CMC), i.e. cholate, CHAPS and β-octylglucoside, over the others, since solubilization usually occurs at or below the CMC where these two detergents due to their higher concentration, relative to the other detergent types, are more effectively solubilizing. However, no single structural property allows one to consider one or the other of the 7 detergent types as being superior in all applications. E.g. a stereochemically rigid hydrophobic tail of cholate or CHAPS is frequently associated with non-denaturing properties and the flexible alkyl tail of SDS or of sulfobetaine detergents with denaturation, but the non-denaturing properties of β-octylglucoside prove that generalization wrong. Similarly, the non-denaturing properties of CHAPS cannot be ascribed to the amphoteric polar head of the detergent since it shares it with the more frequently denaturing sulfobetaines, and since it seems replaceable without apparent ill effects by a carbohydrate moiety (BIGCHAP) [3]. Examples are found in the literature where sulfobetaine [2] or β-octylglucoside are effective non-denaturing detergents while CHAPS is not. The test of 7 detergent types can, of course, only be conducted sequentially. For reasons stated above, it would seem reasonable to start sequential testing with CHAPS and β-octylglucoside.

2.1.3 Detergent Test Conditions

Solubilization depends on the detergent/protein ratio. Therefore, the selected detergent should be tested at a number of concentrations varying between 0.01 to 3%, keeping the protein concentration constant at a high value (5 to 15 mg/mL). Since insolubility is usually not due to hydrophobic bonding alone, as pointed out above, the detergent testing should be carried out in the preferred buffer medium determined as described in Section 2.1.1 . Assay conditions for the protein function may have to be modified under the buffer and detergent conditions of solubilization.

2.1.4 Optimization of Detergent/Protein Ratios

Once a preliminary determination has been made concerning a detergent-buffer milieu in which the protein maintains at least some of its function in the solubilized state, the smallest protein, or protein-lipid complex, capable of maintaining the function (or set of functions) of interest must be determined since this is the species most worthy of biochemical study [2]. Since the size of the solubilized species decreases with increasing detergent/protein ratio, the protein function will diminish as one approaches the smallest functional protein or protein-complex. An obvious limit to decreasing the size of the solubilized species is set by the need to maintain clearly measurable function(s). Having established a preliminary buffer-detergent milieu for maintenance of function in the solubilized state, the investigator has therefore the option of proceeding to finding the conditions for optimal recovery of function independently of the size of the solubilized species (Section 2.1.5), or to finding the smallest functional species (Section 2.2) and to subsequently explore the conditions for maximal activity for that species.

2.1.5   Optimization of the Detergent/Protein Ratio with Regard to Protein Function

Construction of the plot of function *vs* detergent concentration (at constant protein concentration) and of the plot of function *vs* detergent/protein ratio serve to optimize the solubilization medium with regard to function [3]. The first plot will be either sigmoidal or parabolic. If it is sigmoidal, one selects the detergent concentration just sufficient to yield a maximum of solubilized function. If parabolic, one selects the peak value. The selected detergent concentration may then require modification through a systematic variation of the detergent/protein ratio. The minimum detergent/protein ratio yielding maximal retention of function and maximal solubilization is selected.

2.1.6   Comparison Between Solubilized and Insoluble Function

Once a functional protein is solubilized, it is necessary to ascertain the degree to which the original insoluble function has been maintained upon solubilization. This can be done by kinetic assays for enzymes, or by Scatchard plot analysis for receptor proteins. In the latter case, a comparison of slope and x-intercept on the Scatchard plot allows one to determine to what degree the affinity and the number of binding sites have been retained upon solubilization. Further criteria for comparison are parameters such as the maintenance of reversibility and stereospecificity of function [3].

2.2   Determination of the Size of the Solubilized Species

The degree of solubilization and disaggregation of a water-insoluble hydrophobic protein increases in proportion to the detergent/protein ratio. This fact allows one to investigate which the smallest complex, or aggregate carrying the activity or set of activities is, by measuring the protein activity as a function of molecular or solubilized particle size. The gel electrophoretic techniques are singularly adapted for that purpose since they allow one to scan several detergent conditions in a single experiment with a very small protein load requirement compared to chromatographic techniques.

2.2.1   Concentration in a Stack

Protein concentration in a system of sequential moving boundaries, designated as a "stack", is the first gel electrophoretic step toward the elucidation of the size of the solubilized species. The criteria of selection for the required stacking gel, the apparatus and the procedure will be the subject of subsequent Chapters 3 to 9. Here it suffices to point out that 6 to 10 different conditions of the detergent containing gel can be analyzed simultaneously in the apparatus depicted in Fig.6, with protein loads of 5-10 µg/condition when protein staining is used for detection, or less, when detection by protein activity is used. Under the appropriate conditions, the solubilized protein activity appears in a single gel slice which can be eluted for quantitation of the activity. This quantitation comprises not only the particular detergent, detergent concentration or other condition of solubilization established as described in Section 2.1 but also the effect on the protein of electrophoresis in a gel and thus Joule heating, oxidation by polymerization catalysts, effects of concentration in a stack etc. Once a reasonable recovery of the solubilized protein activity is obtained from a particular set of conditions of the stacking gel, the molecular size determination is unproblematic.

2.2.2   Molecular Size Determination

Experimentally, the determination of molecular size by gel electrophoresis requires nothing more than electrophoresis in a single set of gels at various (3 to 7) gel concentrations (Chapter 10). The slope of the resulting linear Ferguson plot provides an assumption-free measure of size. It is only when one wishes to translate this slope into the conventional size parameters, molecular radius and weight, that one needs protein size standards which usually are not available in chemically homogeneous form, standard curves which make assumptions with regard to the geometric configuration of proteins, and needs to make other assumptions which render this translation unreliable. These problems will be fully discussed in Chapter 10. It is also necessary to keep in mind that the detergent bound to the active complex is part of its molecular weight. Thus, Triton X-100 which is micellar in the usual concentration range of 0.02 to 2%, has a micellar weight of 90 000 [3]–[5]. Since the active complex may contain several micellar units of the detergent as well as any number of proteins, it does not seem surprising that molecular weights in the million range have been determined e.g. for the insulin receptor in Triton [1] by use of a standard curve of hydrophilic proteins which do not bind the detergent.

2.2.3   Minimization of the Molecular Size of the Functional Protein or Protein Complex

The gel electrophoretic method for determining molecular size can now be applied comparatively to any number of conditions of detergent concentration, detergent type or mixture of detergents in order to pinpoint the smallest active species. The results of Section 2.1.5 concerning the choice of the minimal detergent/protein ratio may have to be modified if it turns out that a larger detergent/protein ratio, and therefore a higher degree of inactivation, are needed to reduce the size of the functional species. Different detergent types may yield very different sizes of solubilized species. E.g. the size of the native membrane protein, cytochrome P-450, in CHAPS is 100000, while it is more than 300000 (the exclusion limit of the gel filtration column used) in Na-cholate [6]. Similarly, the molecular size of adenylate cyclase analyzed in 0.01% Lubrol PX is twice as much as that analyzed in 0.1% of the same detergent. The size of the larger complex derived from a milieu of 0.01% detergent can also be cut in half in a mixture of 0.1% Lubrol and 0.03% Na-cholate, but not in 0.1% Lubrol alone [2]. Thus, detergent mixing may be an effective tool in the size reduction of active solubilized complexes.

3 Concepts and Tools in the Selection of a Separation pH

The selection of a separation pH is important for the electrophoretic separation of proteins because proteins differ in their net charges as a function of pH across nearly the entire pH-range. Fig.1 illustrates that point on hand of the separation of ovalbumin from trypsin which exhibit representative titration curves (left panel). Separation is best between their isoelectric points of 4.2 and 6.1, respectively, where their net charge ratio is 0 and infinite (right panel) and in the pH-range between those pHs where the two proteins migrate in opposite directions. The selection of a suitable separation pH is for additional reasons important for the separation of cellular particles and of cells, which are viable only within a rather narrow pH-range.

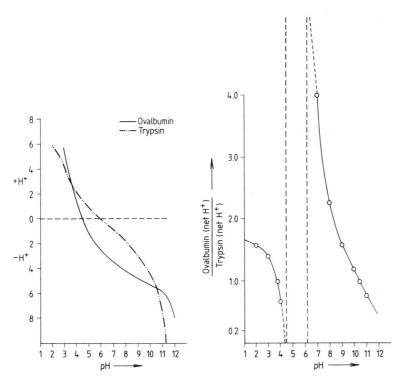

Figure 1. Net charge differences between trypsin and ovalbumin as a function of pH. Titration curves at 0 °C, ionic strength 0.1 M, for both proteins demonstrate the improvement of "charge fractionation" upon approximation of the pIs of 6.2 and 4.4, for trypsin and ovalbumin, respectively. Data of [189] for trypsin, and [190] for ovalbumin.

Separation pH is relatively unimportant in the separation of the SDS-derivatives of the protein subunits, at least in the pH range above neutrality (see Chapter 15), and in the separation of nucleotides. For these reasons, we will designate the species of interest as "the *protein*" in the subsequent discussion, although all considerations equally apply to alternative charged molecular species.

3.1   Electrophoresis in a Continuous Buffer

Assuming that the protein of interest is available in *concentrated* form, e.g. at 1–3 mg/mL or higher, the choice of pH of gel electrophoresis poses no difficulties. Ordinary buffers across the entire pH range are compatible with the polymerization of crosslinked polyacrylamide, and all but the most extreme pHs are compatible with the chemical stability of its amide groups. One will select as buffers weakly dissociating acids and bases in preference to strong acids and bases in order to reduce conductance and Joule heating, and increase voltage across the gel (at a given level of current) and therefore migration rates. For the same reason, buffer concentrations should not exceed 0.01 M if solubility and the maintenance of function of the protein allow. If higher buffer concentrations are needed, as in the above-mentioned case of myosin (soluble in 0.6 M salt), correspondingly long migration times are required in order not to exceed the tolerable level of Joule heating.

When concentrated protein samples are available, thin starting zones can be formed by electrophoresis in a continuous buffer through loading small volumes of sample, and resolution then is the same as that obtained when the starting zone is a "stack" (see below) [7]. Continuous buffers are attractive in gel electrophoresis because the physical milieu in which the protein finds itself is unequivocally defined without any need to calculate or compute the operative parameters as necessary in application of discontinuous buffer systems. Pre-electrophoresis of the gels, to remove the charged components of the polymerization reaction remaining in the gel, is straightforward as compared to the buffer restrictions on pre-electrophoresis prevailing in the case of discontinuous buffer systems. However, even with available concentrated protein samples there are practical advantages to using discontinuous buffers (see below) over continuous ones: The most important is independence from sample volume. Also, with discontinuous buffer systems the progress of electrophoresis can be monitored by a dye zone the width of which remains unaffected by migration distance; i.e., it can be monitored with equal accuracy independently of gel length. For the same reason, the precision of the measurement of migration distance of the dye, upon which determination of the characteristic relative mobility (R_f) value for each zone rests, remains independent of migration distance. The alternative in the use of a continuous buffer is to measure R_fs relative to a marker zone, usually a dye, which spreads progressively with migration distance, rendering the R_f measurement inaccurate and imprecise, or to assign absolute mobilities (cm^2/s/V) to each zone whose determination then requires measurements of migration time (cm/s) and voltage gradient (V/cm) across the gel which are laborious to make compared to an R_f measurement.

3.2   Discontinuous Buffer Systems

The normal situation in biochemistry is that concentrated solutions of the protein of interest are *not* available, at least not without the work and hazards of a separate concentration step prior to separation. The use of *discontinuous* (or multiphasic) buffer systems in gel electrophoresis presents an electrophoretic trick by which the labor and hazard of a pre-concentration step can be entirely avoided, and which automatically provides uniformly concentrated starting zones as well as a procedure of pH optimization (and optimization of other conditions of PAGE) which depends on analysis not of the entire gel for the band of interest, but rather on the analysis of a single gel slice for each condition tested, as will be shown below.

What is the nature of the *electrophoretic trick*?

3.2.1   Moving Boundary

Discontinuous buffer systems are distinguished from continuous ones in electrophoresis by the presence of one or several moving boundaries (Fig.2). Moving boundaries arise from stationary phase boundaries across a relatively rapidly migrating (*leading*) and a relatively slowly migrating (*trailing*) charged species. Conventionally, the initial phase boundary is between a leading ion within a gel, and a trailing ion in the adjacent buffer reservoir. (We will apply the term "ion" indiscriminately to either ions, weak acids and bases, and acidic or basic ampholytes.) This initial, stationary phase boundary moves into the gel, in the direction of migration of the leading ion, once the current is initiated. It is then called a *moving boundary*. Appropriately, that form of electrophoresis giving rise to one or more moving boundaries is called moving boundary electrophoresis (MBE) [8].

3.2.2   Concentration

Qualitatively, MBE provides protein concentration within a moving boundary because the net mobilities and concentrations of ions in the leading phase (the gel prior to electrophoresis) i.e. the leading and common ion define (regulate) the ion

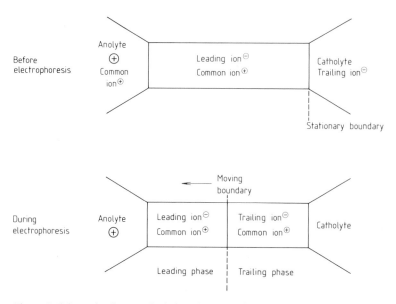

Figure 2. Schematic diagram depicting the generation of a moving boundary in the electric field imposed on a discontinuous buffer system. The discontinuous buffer system consists of a leading ion (with relatively high net mobility) and a trailing ion (with relatively low net mobility) separated by a phase boundary (e.g. a gel/buffer interface). In the electric field, this boundary is displaced in the direction of migration of leading and trailing ions and then designated as a moving boundary. The counterion of leading and trailing ions is identical and thus designated as the common ion of the system. Arbitrarily, a system of negative polarity (leading and trailing ions negatively charged) is depicted.

concentrations in the trailing phase (the gel volume occupied by the trailing ion after electrophoresis is initiated). That regulation by ionic species which exist in the leading phase at concentrations of the order of 0.01 M gives rise to concentrations of the ions in the trailing phase which are somewhat less but of the same order of magnitude (compare columns 3 and 4 with 8 and 9 of the Buffer System Tables of Appendix 1). Considering that the trailing ion may be a protein, regulation by the leading phase of a trailing protein phase to about 0.01 M gives rise to protein concentrations of the order of 100 mg/mL, assuming a MW of 100 000. Protein concentrations within the moving boundary of that order have been measured [7].

3.2.3   Alignment

The protein trailing phase interposed between the original leading and trailing phases serves, in turn, as a leading phase with respect to the original trailing ion, thus regulating a trailing phase of its own. The protein is interposed in a phase of its own between the original leading and trailing ions because it possesses, and only if it possesses, an intermediate "net mobility" (ionic mobility multiplied by the mole frac-

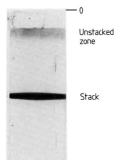

— 0

Unstacked
zone

Stack

Figure 3. The distinctive appearance of a stained protein stack in
contradistinction to an unstacked band.

tion which is ionized) at the given pH between the leading and trailing ions. N
charged molecular species will give rise to N phases at the steady-state, aligned in
order of the net mobilities of the separated constituents (ions, buffer ions, charged
detergents, carrier ampholytes, spacers or proteins), and separated by N-1 moving
boundaries. Such a train of aligned constituents, migrating at the same rate at the
steady-state, can be pictured as a stack of coins, superimposed on another, and migrating unidirectionally in the electric field – hence the term "stack" [9]. It has to be
kept in mind that while macroscopically a stack appears as a single zone with characteristically sharp zone boundaries (Fig.3) it is microscopically a train of all of the
charged constituents present with net mobilities between those of the leading and
trailing ions. This train at the steady-state may contain any number of constituents,
of any chemical type, aligned in order of net mobility.

3.2.4   Mechanism of Moving Boundary Formation

What physical mechanism are regulation of concentration and alignment in order of mobility due to? The answer to that question, as to many others in physics,
can only be provided in mathematical language. The function by which the composition of the trailing phase and the displacement rate of the moving boundary are governed is called the *moving boundary equation.* It is derived in the simplest manner
in application to the widely known Tris-chloride-glycinate buffer system of Ornstein [9] on p.10 of [8]. Qualitatively, the moving boundary equation arises from a
consideration of the ion content of the volume element containing the trailing phase
after passage of a unit of current (the trailing phase in Fig.2). The moving boundary
equation for each constituent is derived from consideration of the necessary bal-
ance within that volume element between mass transport and current transport.
Thus, the mass law dictates that the change in the concentration of an ion is equal to
its electrophoretic transport into or out of the volume element. Secondly, the law of
electroneutrality dictates that for a given transport of charge out of the volume element an equivalent number of charges has to be transported in. Regulation of concentration of the ions constituting the trailing phase is the result of such simultaneous consideration and balancing of mass transport and electrical transport.

3.2.5   Theory and Terminology of Moving Boundary Electrophoresis

Although the derivation of a moving boundary equation for a single constituent is relatively simple [8], the practical elaboration of the concentrations, pHs and conductances of even as simple a MBE system as the Tris-chloride-glycinate system is complex (see p.11 of [8]). Part of the complexity is due to the fact that the mobilities of the leading, trailing and common constituents are only in the case of strong electrolytes (Na^+, Cl^-, SO_4^{2-} etc.) equal to their "ionic mobilities" at a particular ionic strength and temperature. In the case of weak acids and bases, the "net mobilities" enter into the moving boundary equation, i.e. the products between the ionic mobilities and the degree of dissociation at any one pH, as defined by the Henderson-Hasselbalch equation. The moving boundary equations for each ion of the system must be solved simultaneously to account for it in practical physico-chemical terms. The sum of the moving boundary equations for all ions in two adjoining phases leads to an expression known as the "Kohlrausch *regulating function*". This function is, however, not by itself sufficient to calculate the composition of a moving boundary system except when it is composed of strong electrolytes. A further theoretical problem is to ascertain that a given combination of constituents is able to establish a steady-state system of sequential moving boundaries.

To arrive at practical MBE buffer systems, 3 sets of theoretical treatments were developed. Unfortunately, these have used different terminologies, were labeled by different methods' designations and consequently were erroneously considered to be different methods. These are Multiphasic Zone Electrophoresis (Jovin [10]), Isotachophoresis (Everaerts [11]) and Steady-State Electrophoresis (Schafer-Nielsen and Svendsen [12]). Less comprehensive treatments used the names of Steady-State Stacking (Ornstein), Omegaphoresis (Schumacher and Ryser) and Displacement Electrophoresis (Martin, Hjerten) for what is now known to be a single identical method i.e. moving boundary electrophoresis. Most treatments rest on the original theoretical formulations of Svensson (Rilbe) and Longsworth and coworkers worked out between 1943 and 1953 (see [8]).

For the practice of MBE, neither the multiplicity of methods' labels nor the complexity of theory matter. The former turns out to be an airbubble which exploded once the single identity of MBE and the interconvertibility of its languages had been demonstrated [8]. The second is irrelevant because the theoretical treatments of Jovin, Everaerts and Schafer-Nielsen have been incorporated into computer programs which are easy to use.

3.2.6   The "Extensive Buffer Systems Output"

Of the 3 programs, it is that by Jovin [13] which is the most useful in practice since its output is numerical, explicit, and detailed, providing an exhaustive description of the physical properties of each of the buffer phases as well as numerical ways to modify leading and trailing ion mobilities which define the mobility range of the proteins which can be stacked.

The computer program of Jovin [13] has been used to generate a) a model computer output of several thousand exhaustively described buffer systems ([14] and Appendix 2) and b) a selection of 19 buffer systems presented in a simplified terminology and format and limited to some essential parameters for the practice of gel electrophoresis ([15] and Appendix 1).

3.2.6.1   The extensive output

The computer output of buffer systems capable of stacking proteins at any pH is available at a nominal price from the National Technical Information Service, US Department of Commerce, Springfield VA 22151, USA, in microfiche form, under the "Public Board (PB) Numbers" given in Table 1, Appendix 2. A limited number of such buffer systems has been printed in published reports (Table 1, Appendix 1). In addition, the major empirically derived discontinuous buffer systems have been analyzed by use of the Jovin program with regard to their operative properties [16]. The buffer systems of the extensive output have been computed at 1/2 pH unit intervals, for negatively charged proteins (negative polarity) from pH 5 to 11, and for positively charged proteins (positive polarity) from pH 2.5 to 9.0. The resulting 4269 buffer systems have been catalogued. The catalog (for PB number, see above) lists leading and trailing constituents, the relative mobilities of leading and trailing ions, the pHs of leading and trailing phases for each system and indicates ranges by which pHs and mobilities can be varied for the particular assortment of constituents of a given system. The buffer systems output itself appears in 3 pages for each system, where the first page reiterates the input parameters (not shown), the second provides a physico-chemical description of each of the stacking and resolving buffer phases, including pHs and relative mobilities, and a recipe for preparing stacking and resolving gel buffers and the catholyte (negative polarity) or anolyte (positive polarity). The 3rd page describes the "subsystems" discussed in the following section. A detailed guide to the use of these systems has been reported previously [17],[18], and is here provided in Appendix 2.

3.2.6.2   Selected discontinuous buffer systems

The selected buffer systems output presented in a simplified *terminology* and format avoids the complexities of the Extensive Buffer Systems Output (Section 3.2.6.1). The 19 buffer systems of that selection [15] (Appendix 1) provide discontinuous buffer systems across the *pH-scale* at 0.01 ionic strength and 0 °C operative at approximately 0.5 pH unit intervals. Since *proteins* exhibit low net mobilities even in "non-restrictive" gels compared to buffer ions, these buffer systems have been selected to provide a low trailing ion net mobility of approximately 0.050 (relative to the mobility of Na$^+$ at the same temperature and ionic strength). Most proteins exhibit a net mobility of 0.050 at least at the extremes of pH and often at more neutral pH values and can therefore be stacked in such a buffer system. An explicit buffer *recipe* is provided. Finally, the 19 buffer systems' tables specify the buffer "subsystems" with trailing ion net mobilities *increased* from 0.050 to 0.250. Such an increase serves various functions: Unstacking in a resolving gel to be able to separate proteins according to their restricted mobilities in a restrictive gel more easily; specific stacking for the protein of interest, providing separation in the stacking gel prior to entrance into a resolving gel; regulation of R_f values in the resolving gel to obtain values between 0.25 and 0.85 at a desired gel concentration. All of these features of the selected buffer systems will be discussed in detail below.

The *terminology* of the buffer systems tables (Fig.4 and Appendix 1) is that used in Fig.2 to describe a moving boundary. The leading phase consists of leading and common ions, the trailing phase of trailing and common ions. The term trailing ion mobility denotes the *net* mobility of the trailing ion relative to the mobility of Na^+. It approximates (see below) the lowest net mobility the protein has to attain at the pH of the trailing phase to be able to migrate within the system of sequential moving boundaries (the stack). The symbol "4 x" in the recipe section of the tables signifies 4-fold the final concentration in the gel (for reasons outlined below). In a buffer system with negative polarity (leading and trailing ions negatively charged as in Fig.2), the catholyte is a reservoir for trailing ion, the anolyte a reservoir for common ion; in both cases, the choice of counterion or pH is irrelevant to the moving boundary system. In a buffer system with positive polarity, the anolyte is the reservoir for trailing ion.

Unspecified features common to the buffer systems of Appendix 1 are that trailing phases and therefore, operative conditions for the protein, are given at 0.01 ionic strength, 0 °C. Values of pH throughout the tables have been corrected for the temperature and ionic strength of the buffer to whose molar composition or mode of preparation the pH is assigned.

The *organization* of the buffers system tables of Appendix 1 is exemplified by Fig.4. Column 1 shows the operative pHs by which an initial selection of a buffer system is made. The pH of the trailing phase is "operative" with regard to the protein in as far as it approximates the pH of the "protein boundary" sandwiched between the leading and trailing phases. (Since protein mobilities are usually much closer to trailing ion than to leading ion mobilities, this approximation appears justified). The 2nd column is probably the most important in practice. It allows us to vary trailing ion mobilities at will between values of about 0.050 and 0.250, by preparing the gel buffers specified in the same line as the mobility, using columns 5 and 6. Any buffers giving intermediate mobility values can be obtained by graphic interpolation. Thus, any protein with a mobility larger than 0.050, and less than 0.250 can be stacked by use of column 2. Equally, any protein with mobility less than the trailing ion mobility can be unstacked in a resolving gel. Stacking can be made specific for the protein by increasing the trailing ion mobility in column 2 and using the corresponding gel buffer.

Columns 3 to 7 give the composition of the gel buffer, both in molar concentration (for documentation) and in terms of a buffer recipe. The recipe refers to a 4-fold concentrate of the gel buffer since polyacrylamide and agarose gels can be conveniently prepared by such a concentrate (see Chapters 5,6). The recipe also refers to 25 °C, to provide ease of measurement, although the operative temperature in all buffer systems given in Appendix 1 is 0 °C. The pH in column 7 is adjusted to both 25 °C and the molarity of the 4-fold concentrated gel buffer.

| | | | Catholyte | 4.58 g TES + 10.0 mL 1 N KOH/L; $pH_{25°C}$ = 7.44 |
| Anolyte | 4.18 g Bistris + 10.0 mL 1 N HCl/L; $pH_{25°C}$ = 6.50 |

Trailing Phase $pH_{0°C}$	Trailing Ion Mobility	Leading Phase						Trailing Phase		
		Leading Ion Cacodylic acid (mol/L)	Common Ion Bistris	4× Leading Ion Cacodylic acid (g/100 mL)	4× Common Ion Bistris	$pH_{25°C}$		Trailing Ion TES (mol/L)	Common Ion Bistris	$pH_{25°C}$
7.05	0.045	0.1016	0.0358	5.61	3.00	5.78		0.0905	0.0247	6.69
7.39	0.086	0.0530	0.0477	2.93	3.99	6.29		0.0472	0.0419	6.94
7.61	0.127	0.0360	0.0677	1.99	5.67	6.71		0.0321	0.0638	7.17
8.15	0.250	0.0163	0.1972	1.01	16.67	7.61		0.0183	0.1992	7.74

Buffer System Number 12

Figure 4. Representative discontinuous buffer system suitable for moving boundary electrophoresis of proteins at 0 °C, 0.01 ionic strength [15]. For buffer systems operative at other pH values, see Appendix 1. The symbol "4x" signifies 4-fold the concentration over that in the gel.

Columns 8–10 give the composition of the trailing phase, both for documentation of the operative conditions and for the purpose of testing the stability of a protein function in the particular operative buffer milieu, prior to any electrophoresis. Again, for ease of buffer preparation in the latter case, the pH is given at 25 °C.

Columns 11 and 12 give the recipe for preparation of the buffers contained in the cathodic and anodic buffer reservoirs. In apparatus with vertical geometry, the catholyte is the "upper buffer", the anolyte the "lower buffer" for systems with negative polarity (Appendix 1, Tables 1–14). Conversely, for systems with positive polarity (Appendix 1, Tables 15–19), the anolyte is the upper buffer.

3.2.7   Alternative Formulations of Moving Boundary Electrophoresis

An alternative approach to the "Extensive Buffer Systems Output" for the systematic stacking of proteins is that of Schafer-Nielsen and Svendsen [12] designated as Steady-State Electrophoresis and computerized in a form providing a graphic output for the net mobility of each buffer constituent *vs* pH. This should enable one to read off the prepared and the operative buffer concentrations given numerically in the Extensive Buffer Systems Output, but to date no documentation which would allow one to do that has been published, so that our discussion will be restricted to the practical use of the competing Extensive Buffer Systems Output. Similarly, the older program of Routs [19] which embodies the theoretical treatment of moving boundaries by Everaerts [11] does not provide the user with explicit trailing phase compositions and properties, nor does it provide recipes for preparation of leading phase buffers. Both the Schafer-Nielsen and the Routs programs deal, however, with the setting up of sequential multiple moving boundaries (although without testing for their steady-state [8]), while the Extensive Buffer Systems Output confines itself to single moving boundaries. A recent application of the Jovin theory to near-stationary sequential multiple moving boundaries [20], however, shows that this is not an inherent limitation of the theory. A unique feature of the Schafer-Nielsen treatment is the consideration of highly mobile species in the trailing phase. It remains to be seen to what degree this might provide practical advantages.

After defining the solvent for separation (Chapter 2), and adopting discontinuous buffer systems (Chapter 3) for reasons of sample concentration, experimental ease in pH optimization and the possibility to conduct pH optimization in a manner specific for the protein of interest through selection of narrow ranges between leading and trailing ion mobilities we would be ready to proceed with a step-by-step display of the experimental procedure for pH optimization, if it were not for the fact, that the armory of equipment needed for pH optimization as well as for the later pore size optimization and resolution under optimal conditions, remains undefined, as is the gel matrix of polyacrylamide or of agarose upon which many of our experimental procedures depend. We will therefore interpose a discussion of the apparatus needed to do Quantitative PAGE (Chapter 4), and a discussion of polyacrylamide and agarose gels (Chapters 5, 6), before proceeding with the experimental definition of the optimal separation pH (Chapters 7,8).

4   The Apparatus of Quantitative PAGE

Quantitative PAGE rests on the reproducibility of protein mobilities in gels. Thus, apparatus is required which through control of electric field strength, and through control of temperature provides reproducible free electrophoretic mobilities, and in addition apparatus is needed which allows us to make reproducible gels through control of the conditions of the polymerization reaction such as temperature, oxygen inhibition and photocatalysis. The discussion of the design elements of the apparatus will allow the reader to discriminate between commercially available apparatus types, and to realize its present shortcomings. Apparatus constructed in the author's workshop will be presented in order to show how some of these shortcomings can be overcome. This will benefit directly those in command of a workshop, and allow those who are not to specify their demand from commercial sources. Representative commercial forms of apparatus from the manufacturers and distributors listed in Table 1 of Appendix 4 will be cited. These citations reflect the necessarily limited experience of the author with commercial equipment and serve to provide the reader with examples of the various design types. The list does not in any way represent a preference for a cited source over others that may not have been cited.

4.1   Temperature Control Devices for Polymerization and Electrophoresis

Electrophoretic mobilities vary steeply with temperature. Reproducibility of mobilities therefore depends on temperature control. Another important reason for strict temperature control and maximum heat exchange rates between gel and apparatus lies in the fact that resolution improves in direct proportion to voltage, making it desirable to operate in electrophoresis at the maximum voltage compatible with the dissipation of Joule heat. Finally, temperature control is required in the electrophoresis of native proteins at 0–4 °C to preserve protein activities as much as possible. If polymerization of polyacrylamide is carried out in the same apparatus as electrophoresis, as advocated here, another rationale for temperature controlled apparatus arises from the need to dissipate the heat of polymerization efficiently. In PAGE, 0 °C operation is preferred even in cases where activity is of no concern since pore size reproducibility is better at 0 than at 25 °C (Chapter 5). Also crosslinking agents participating in the polymerization reaction relatively sluggishly, like N,N'-diallyltartardiamide (DATD), give rise to more homogeneous gels at 0 °C than at ambient temperature, and in some cases gel only at 0 °C. Agarose electrophoresis requires stringent temperature control in its quantitative application [40],[22] since gelation rates and resulting pore sizes vary with temperature, and in view of the melting of gels at temperatures that may be obtained as a consequence of Joule heat during electrophoresis, particularly when low-melting agaroses are used [21].

Since liquids are more efficient in heat dissipation than gases, it is better, from the viewpoint of achieving maximum rates of heat transfer, to cool the gel to the desired temperature by means of circulating liquid coolants than by conducting gel electrophoresis in the cold room. An adequate degree of temperature control of polymerization and electrophoresis is most easily achieved by a) carrying out both in water-jacketed apparatus; b) magnetic stirring of the lower buffer during polymerization and electrophoresis; c) maintaining a coolant flow rate of 1 L/min or more. Coolant should contain enough ethyleneglycol or glycerol to prevent freezing. Refrigerated water baths maintained a few degrees below the desired temperature in the gel electrophoresis apparatus are required. Since maintenance of 0 or 25 °C, and not temperature control within narrow tolerances at *any* temperature is required, the usual refrigerated water baths (e.g. Brinkmann, Lauda K-2/R) are unnecessarily versatile and expensive for the purposes of gel electrophoresis. Refrigerated fraction collectors which employ liquid coolant at 0 °C are suitable for our purpose (e.g. Buchler No.3R-40002 T, presently discontinued), but involve the unnecessary expense of a fraction collector and are excessively bulky. A most precise but costly way of cooling is through a temperature control unit (Hoefer No.TC 380) immersed in the gel electrophoresis apparatus which activates an electronic cold finger (CryoCool, Neslab No.CC-60,80,100) in a bath surrounding the lower buffer reservoir as needed to maintain 0 °C in the apparatus. However, the much simpler design of a water bath capable of control at exclusively 0 °C and 25 °C (i.e. operating 1–2 degrees below those temperatures) would appear more economical and more stable. At this time, this is not as yet commercially available.

By far the most efficient, although also the most expensive, presently known method of temperature control is electronic, and exploits the Peltier effect. It has been applied to date only in 2 applications, the horizontal gel slab apparatus (MRA No. M 157, Hoefer No. HE 950, Isolab No. FR-2100) and the free-flow Hannig apparatus (Desaga, Brinkmann No.FF-3, discontinued). An important application of the horizontal slab apparatus has been to ultrathin gels electrofocused at voltages as high as 800 V/cm with a corresponding improvement of resolution [23]. It is to be expected, in spite of these modest beginnings, that the future development of apparatus for electrophoresis at maximum tolerable voltage to provide maximum resolution should incorporate cooling by Peltier effect, in particular in conjunction with intermittent periods of cooling between high voltage pulses of power and Joule heating of the gel (see Section 4.10).

4.2 Gel Tube Apparatus

Certain types of experiments define the particular usefulness of gel tube apparatus and its place in the armory required for gel electrophoresis, just as the other apparatus geometries discussed below have well circumscribed areas of usefulness of their own. The notion that a single apparatus type could replace all the others in all applications is on its face unscientific and a product of wishful thinking on the part of manufacturers anxious to promote any apparatus type which can be inexpensively produced and most profitably sold.

Gel *tube* apparatus is specifically useful whenever PAGE is carried out with gels of *various gel concentrations* in the same experiment, as in the construction of Ferguson plots, or when gel electrophoresis is to be begun at the same times, but terminated at *various times* for different tubes, as in the timed gel electrofocusing experiment. (Note that the latter differs from a slab gel electrofocusing experiment in which samples are loaded onto the various channels at different times, since in that case, each sample is loaded onto a pH gradient at a different point of the life cycle of that gradient. In electrofocusing using Immobiline, this may not be the case.) Thirdly, gel tube apparatus is to be preferred over gel slab apparatus when the sample *volume* is large, i.e. larger than usually available in the sample slots of gel slabs which rarely exceed a capacity of 50–100 µ L. Finally, when only one or a few samples are to be run, it is more *economical* to use a single tube than to prepare a gel slab suited for many samples.

Rudimentary tube apparatus consists of a 1 or 2-hole rubber stopper topped by a section of glass tubing of sufficiently wide diameter (the upper buffer reservoir); the glass tubes held by the stopper; and a beaker into which the glass tubes reach (the lower buffer reservoir). Ornstein and Davis developed this rudimentary model to the basic type still used today [24], by replacing the rubber stopper with rubber grommets and by increasing the size of the upper buffer reservoir to hold 10–14 tubes. Contemporary models have added a few additional design elements to the Ornstein-Davis model, of which additions temperature control is the most important.

At least four types of contemporary, commercially available gel tube apparatus must be distinguished. Model I (Fig.5) is that used in the author's laboratory. It consists mostly of Pyrex and is constructed by Pyrex fusion. Blueprints for this model are available from the Biomedical Engineering and Instrumentation Branch, NIH, Bethesda MD 20205 under No. 71-73-103 and 71-73-103M. A similar model is available from R & D Scientific Glass Co. Model II is similar in design to I except for the cooling system and the fact that it is molded from a fluorohydrocarbon, a highly chemically resistant plastic. It is produced by Hoefer (No. GT-3,6,14). Model III is another design similar to I, but restricted to use with 5–6 mm diameter gels. It is produced by Miles (No. 43-510-1). Model IV is the "Polyanalyst" apparatus of Buchler (No. 3-1751A), which combines a Plexiglas upper buffer reservoir with a glass lower buffer reservoir, and Model No. DE-102 of Hoefer which is an all-Plexiglas construction. The quality and chemical resistance of the Plexiglas is better in the Hoefer than in the Buchler apparatus. Both feature uncooled upper buffer reservoirs rest-

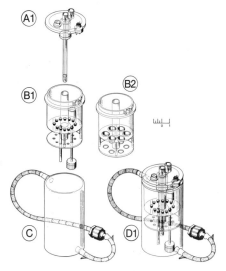

Figure 5. Gel electrophoresis tube apparatus Model I. Apparatus materials are Pyrex unless indicated otherwise. (A1) Electrode top. Male banana plugs, male safety interlock component and spirit level are mounted on a polycarbonate plate. (B1) Upper buffer reservoir for 6 mm tubes, with or without band collection cup. Tubes are seated in rubber grommets and aligned vertically by the bottom alignment plate. The sleeve is made of polycarbonate. (B2) Same, for 18 mm tubes. (C) Jacketed lower buffer reservoir. (D1) Assembly. Glass-glass interfaces are bonded by Pyrex fusion; glass-plastic interfaces by RTV (General Electric). Construction details and commercial source are given in Chapter 4, together with a detailed discussion of the design elements of the apparatus. Dimensions are given in inches.

ing on top of the water-jacketed lower buffer reservoir. Thus, they do not provide hydrostatic equilibration of the gel. The cost of Model I is the highest, of Model IV the lowest.

The following design features for gel tube apparatus will be considered:

- Temperature control
- Hydrostatic equilibration of the gel
- Applicability to various tube diameters and geometries as well as different gel types
- Vertical alignment
- Condensate drainage
- Safety interlocks
- Apparatus materials
- Positioning of the tubes in the upper buffer reservoir
- Possibility to remove gel tubes individually

4.2.1   Temperature Control

In common with all other types of electrophoresis apparatus is the need for temperature control, providing efficient dissipation of the heat of polymerization during formation of the gel and of Joule heat during electrophoresis. Models I, III and IV provide a *jacketed* lower buffer reservoir, through which coolant is pumped at a rate of 1 L/min or more. If the coolant reservoir is maintained 1–2 °C below the desired gel temperature, this provides adequate temperature control within gel tubes fully immersed in that reservoir, provided that the lower buffer is continuously circulated by magnetic stirring. In Model II, where the reservoir is cooled by a central *cold finger*, the bath temperature must be maintained at -7 to -9 °C to maintain 0–2 °C in the gel. This problem may be resolved through compensating for the low cooling surface of a cold finger (compared to a jacket) by placing a thermal control unit (Hoefer No. TC 380) into the apparatus and allowing it to activate an electronically refrigerated cold finger (Neslab Cryocool) in a water bath surrounding the apparatus.

The upper limit of gel tube *diameter* compatible with temperature control at the center of the gel within 0.1 °C during electrophoresis appears to be 15 to 18 mm, when realistically rapid rates of electrophoretic migration (in hours, not days) are to be maintained.

It is obvious that temperature control becomes easier as the tube diameter – or the thickness of the gel of any geometry – is decreased. Correspondingly, thin gels allow one to apply high voltages which improve resolution. This approach to good temperature control will be discussed at length in the section on horizontal slab apparatus (Section 4.4). Here, it suffices to note that miniaturization of gels runs into problems associated with a large surface to mass ratio and with surface effects on the gel. Thus, the protein mobilities obtained in 5 μL capillary tubes are much higher in nominally identical gels than in a 6 mm ID tube (compare the albumin mobility in [25] with that observed in 5–6 mm tubes in the Ornstein-Davis buffer system). Miniaturization also brings about an undesirable degree of artfulness in the manipulation of the gels.

Unlike in horizontal slab apparatus (see below), electronic cooling based on the Peltier effect has not been used to date in the design of gel tube apparatus, although it can be expected to provide the possibility to do Quantitative PAGE at much higher voltages, migration rates and resolving power than is presently possible.

4.2.2   Hydrostatic Equilibration of the Gel

This term implies the absence of hydrostatic pressure on the gel by the weight of the upper buffer. Equilibration is brought about by immersing the upper buffer reservoir into the lower one and by maintaining *equal* fluid *levels* in both. It is of importance whenever wall adherence of the gel to the tube is weak, i.e. in application to open-pore gels like the "non-restrictive" gels in MBE, and to EF gels which are mechanically labile presumably because of differential gel swelling along the gel length as a function of pH and local voltage maxima (Section 12.2). Models I, II and III but not IV provide hydrostatic equilibration.

4.2.3   Applicability to Various Gel Diameters and Geometries as well as Gel Types

Gel tube apparatus should be versatile in lending itself to both polymerization and electrophoresis in conjunction with different gel types, as well as in allowing for various tube diameters, tube lengths and other gel geometries in order to make it applicable to a variety of separation problems. For the usual analytical scale PAGE, 5 or 6 mm diameter gels seem the most convenient to use. When the sample is scarce, load can be cut to 1/4 by conducting PAGE in tubes of 3 mm ID. Wall surface effects on polymerization (see above) cannot be observed with 3 mm ID gel tubes. Gel removal is considerably easier with 3 mm compared to 2 mm ID gels. Preparative PAGE requires a maximal gel diameter compatible with adequate dissipation rates of the heat of polymerization and of Joule heat at practical rates of electrophoresis, i.e. tubes of 15 to 18 mm ID. Since heat transfer from the tube to the lower buffer is crucial, wall thicknesses exceeding 1 mm should be avoided. Tube apparatus should thus provide *exchangeable* upper buffer *reservoirs* for use with 3, 6 and 18 mm diameter gels or similar sizes. Models I and II do provide those, but not III or IV.

Gel tubes in those dimensions made of Pyrex (see Chapter 5) should be used disposably. They are best pre-cut to the desired length from standard length tubing (e.g. Thomas, No.5692-E30), lightly fire-polished and stored dustfree in a closed container. If the gel concentration is too low or if gel swelling along the gel length too uneven to provide sufficient mechanical support by merely balancing the weight of the gel through the adherence of the gel to the tube walls, several or all of the following measures for strengthening the adherence of the gel to the wall should be taken. a) Tubes should be cleaned with dichromate *cleaning* solution or alcoholic KOH and rinsed to neutrality. b) Tubes should be coated with 1% linear polyacrylamide (Gelamide 250, Polysciences, No.5883) or with 1% electroendosmosis-free agarose (Iso-Gel, Marine Colloids). In each case, the procedure of *coating* consists of filling the gel tube with the 1% solution, and allowing it to drain immediately onto adsorptive tissue. In the case of agarose, draining of the liquified polymer proceeds in an oven at 60 °C. Drying time is 1 h at 60 °C for agarose, overnight at R.T. for Gelamide 250. The conceptually most attractive way to provide firm adherence of the gel to glass walls is to bond it covalently. This can be done by coating the glass walls with a siliconating agent linked to a methacryl group (Silane A-174, Pharmacia No.17-0466-01). This methacryl group is incorporated (although at a low rate, see Chapter 5) into the growing polyacrylamide chain during polymerization, thus providing a covalent bond between gel and glass. In practice, however, removal of such gels from the tube is very difficult, making this mode of strenghtening wall adherence inapplicable to gel tubes, except for the MBE gels used in preparative slice extraction (Section 13.1). The coating procedure consists of filling the tube with 1% Silane A-174 in acetone, allowing to stand for 0.5–1 h, and drying at room temperature. The tubes are then rinsed with water and stored in the cold.c) Mechanical *support* of the gel tubes by Nylon mesh (100 μ mesh size is suitable). After polymerization of the gel, 1–2 cm^2 pieces of the mesh are wetted, slipped over the bottom of the gel tube and fastened there by a sleeve of 8 mm ID Tygon tubing (about 3 mm length). Air bubbles between the Nylon mesh and the gel must be removed before using the gels, by repetitively applying and withdrawing buffer with a small syringe (without needle).

Gel tube apparatus with exchangeable upper buffer reservoirs is also suitable for other than cylindrical gel geometries. In conjunction with nanogram scale protein loads and silver staining rectangular cross-sectional gel tubes (2 × 4 mm) have been used in such an upper buffer reservoir fitting a Model I apparatus [71]. For use in the Model II apparatus, an upper buffer reservoir holding 3 prefabricated vertical gel slabs (IsoLab No.GA-70) is available (Hoefer No. UBS3).

4.2.4 Vertical Alignment

To obtain straight bands, gel tubes have to be vertically aligned. To monitor and achieve this alignment, 3 design elements are required: A spirit level mounted to the electrode top; an alignment plate attached permanently to the upper buffer reservoir providing 2-point support for each gel tube; adjustment screws in the apparatus base by which the apparatus is aligned until the air bubble in the spirit level is centered. Only Model I possesses all of those features.

4.2.5 Condensate Drainage

Condensate drainage is important in humid climates where operation of a gel tube apparatus at 0 °C tends to flood the lab bench, which is inconvenient and possibly a safety hazard during electrophoresis. Only Model I provides an apparatus stand with provision for condensate drainage.

4.2.6 Safety Interlocks

Safety provisions in gel tube apparatus vary. Model I has magnetic safety interlocks between electrode top and upper buffer reservoir which upon use tend to get corroded and then are frequently shorted out in routine use. But safety interlocks of this type have the advantage that they do not interfere with removal of gel tubes during a run by requiring one to disconnect any electrical or cooling lines. In all other models one has to disconnect the leads either from the electrode top or the power supply to remove a gel tube. In Model II it is even necessary to disconnect the coolant lines before a tube can be removed. Since the key applications of tube apparatus involve intermittent removal of gel tubes during a run, all such safety devices requiring severence of lines are highly unsatisfactory and frequently tempt the user into operating without an apparatus cover, i.e. without any safety measures.

4.2.7   Apparatus Materials

Upon prolonged use, Plexiglas apparatus cracks, particularly when the coolant contains anti-freeze such as ethyleneglycol (which it must contain for operation at 0 °C). Also, any glass to plastic bond (with the notable exception of the preparative Buchler Polyprep apparatus), or even plastic-to-plastic bond made by use of a cement is likely sooner or later to give rise to leakage. Therefore, tube apparatus construction by either fluorohydrocarbon plastic molding (Model II) or Pyrex fusion (Model I) is to be preferred. Unfortunately, both processes are expensive. One should also consider the heat conduction properties and the transparency of materials. Transparency is needed for photopolymerization. From a heat transfer viewpoint, the "cold finger" or the inner walls of a waterjacket should be highly heat-conductive thin glass or even plastic-coated metal, while the outer apparatus walls should be thick plastic to provide good heat insulation. Clearly, therefore, both Models I and II suffer from lack of outer insulation, and Model II in addition from poor heat transfer across the PVC-cooling finger.

The choice of tube materials is governed by consideration of the adherence of gels to the tube walls and of heat transfer characteristics. Both polyacrylamide and agarose adhere to plastic tubes if they are coated with a heat-dried film of agarose (as on the precoated plastic sheets used for horizontal apparatus). However, plastic walls are inferior to glass walls with regard to heat transfer. Therefore, glass tubes with as little wall thickness as practical from the viewpoint of fragility (1 mm) are preferable. Since soft glass surfaces are either inhibitory to polymerization or catalyze the reaction to form short-chain polymers (see Chapter 5), Pyrex is the preferable glass to use.

4.2.8   Positioning of the Tubes in the Upper Buffer Reservoir

In Models I and IV (Buchler) gel tubes are held in the upper buffer reservoir by several mm thick rubber grommets spanning across reservoir bottom plates of equal thickness. These are preferable to the thin grommets in Models II and III, necessitated by their relatively thin reservoir bottom plate (apparently it is difficult to mold thick plates), since thick grommets prevent pushing the grommet out of the plate when a gel tube is inserted or withdrawn. Grommets are painful to maintain: When new, they tend to be so tight that tube insertion and withdrawal is difficult. When worn, they tend to leak. Where very tight seals between catholyte and anolyte are required as in EF, grommets should be sealed into the bottom plate of the upper reservoir by RTV-adhesive.

Alternatively to grommets, 1-hole stoppers can be used to hold the tubes as in Model IV (Hoefer). They of course are easier to exchange after some wear than the grommets, but they also are less reliably leakproof and do not provide vertical tube alignment, thus requiring narrow tolerances for a vertical alignment plate (see above) which are inconvenient since outer tube diameters are variable.

Another alternative to grommets are Swagelok-fittings made of a chemically resistant plastic [26]. But individual tightening of these fittings for each tube is too laborious, particularly since they are seated at the bottom of the upper reservoir and therefore hard to reach. It is clear, therefore, that no fully satisfactory way of holding gel tubes in the upper buffer reservoir exists to date. Thick rubber grommets seem the least objectionable solution at this time.

4.2.9   Possibility to Remove Gel Tubes Individually

One of the key advantages of gel tubes over (non-partitioned) gel slabs is the possibility to withdraw selective tubes individually from electrophoretic runs, either in PAGE at different gel concentrations or in gel EF at various focusing times. Such individual tube withdrawal requires an apparatus design such that neither electrical nor coolant connections need to be severed during removal of a tube from the apparatus (see Section 4.2.6). In such tube apparatus types as Hoefer No. GT-5 and MRA No. M91 or M137-LP an individual tube cannot be removed from the apparatus at all without dumping of the buffers and, in the former case, disassembling the apparatus itself.

For intermittent individual tube removal during electrophoresis rubber stoppers fitting the top of the tube (e.g. Thomas No. 8751-K10 for 6 mm ID tubes) are required (see Section 5.5 for the procedure).

4.3   Vertical Gel Slab Apparatus

Slab apparatus in general has its advantages, drawbacks and areas of specific applicability that need to be defined. Within each area of application, separation problems specifically appropriate for *vertical* as opposed to horizontal gel electrophoresis apparatus can be identified.

4.3.1   Applications of Slab Gels in General

Gel slab apparatus is preferable to tube apparatus a) when a comparative analysis of many samples at a *single* gel concentration in PAGE, or at a *single* timepoint in EF is carried out, assuming the sample size to be small (less than 50–100 µL), and b) when a multicomponent system is being analyzed. The latter application of slabs for electrophoresis in 2 dimensions takes advantage of the geometric fact that on statistical grounds, maximally about 50 bands can be resolved 1-dimensionally along a conventional migration path under ideal conditions [27], and thus a square gel of equal length is capable of resolving $50 \times 50 = 2500$ spots (see Chapter 16).

4.3.1.1   Identity testing

A classical application for gel slab apparatus is in the identity testing by visual comparison of migration distances between samples on the same gel, which avoids both the error due to variability in the polymerization reaction carried out in different gel tubes, and the measurement error involved in assigning a characteristic mobility value to the band in each gel tube [28]. An extension of this comparative technique to multiple samples in population genetic studies, and a refinement of resolution achieved by re-electrophoresis of fractions on gel slabs, is exemplified by the "double 1-dimensional" (D1-D) technique [29].

Of great importance to present day biochemistry is the application of slab apparatus to SDS-PAGE. As will be shown in Chapter 15, derivatization and subunit dissociation of proteins by the detergent, SDS, allows one to estimate the molecular weight of a protein on the basis of migration distance at a single gel concentration (within limitations spelled out in the later chapter). Thus, the experimentally simplest way to physically characterize the polypeptide components in a mixture is to assign to each component SDS-derivative a MW value, based on migration distance relative to a series of MW standards run on the same gel. This could be done equally well on a vertical and a horizontal gel slab, if protein samples are concentrated enough so as not to exceed the sample volume capacity of the apparatus. This capacity is lower for horizontal than vertical gel slabs. This fact, together with the compatibility of vertical slabs with large buffer chambers without recourse to wicks, and uniformity of migration across the thickness of the gel, accounts presumably for the popularity of vertical slabs in SDS-PAGE as compared to horizontal ones.

4.3.1.2   Multicomponent system

The application of slab apparatus to multicomponent systems consisting of more than 50 components lends itself to vertical, in preference to horizontal, gel slab apparatus for the same reasons as discussed for SDS-PAGE, since conventionally the second dimension in 2-D-PAGE is SDS-PAGE. Parallel stacking and resolving gel surfaces can, however, also be formed in the horizontal technique by polymerizing the gels with the plates in vertical position, and thus the possibility exists also for horizontally run slabs to embed the 1-D gel in a 2-D stacking gel, using SDS-PAGE in a discontinuous buffer system, with consequent concentration of the 2-D starting zone (Chapter 16).

Although gel slabs possess these areas of specific applicability either to the identity testing, or to the resolution of SDS-subunits at a constant gel concentration, or to EF at a constant focusing time, these applications do depend, nonetheless, on prior optimization of gel concentration or of EF time by experimentation with variable gel concentrations and variable EF times, and therefore on prior use of gel tube apparatus.

4.3.2 Small and Large Forms

Vertical slab apparatus exists in *two forms*: A relatively *small* disposable slab held by rubber grommets in an upper buffer reservoir exactly in the manner of gel tubes (e.g. IsoLab No.GA-70, or Hoefer No.UBS3), or a relatively *large* design requiring assembly of a reusable gel slab and providing spacers to give variable gel thicknesses (e.g. Hoefer No.SE-600-3.0, 1.0, 0.75; IsoLab No.EP-2100; BioRad No. 165-1420). To date, the latter type has been used nearly exclusively.

The attraction of the *small* apparatus is that slabs are prefabricated and disposable and do not need to be assembled; and that they require no special apparatus, since they can be positioned in an exchangeable upper buffer reservoir of tube apparatus Models I and II.

For the identity testing in cases where closeness in mobility between species interferes with distinction by Ferguson plot ("ellipse") criteria (Chapter 10), the small slabs appear convenient. They also appear most suitable for running up to 8 MW standards per gel concentration in PAGE. Thus, with 3 slabs per upper buffer reservoir, one can construct an adequate standard curve in 2 runs. The smaller slab inserts into rubber grommets also seem adequate in migration path for many problems in SDS-PAGE which are limited to one or a few samples. The slab is inserted into the upper buffer reservoir insert in exactly the manner of a gel tube, and is held in a leak-proof, rubber grommet with knobbed 2-point rubber seal (Hoefer). The reservoir of the Model II apparatus holds 3 slabs. Relatively minor items like slot formers of various sizes, a device for sealing the bottom of the slab after introduction of the polymerization mixture, or slabs of various thickness, are not available as yet for this apparatus. Also, at present the disposable slabs (Iso-Lab) are made of soft glass, not Pyrex.

The *large* vertical slab has the advantage that 40 or more samples can be run in the same apparatus at the same time. Its large surface area is required for the resolution of multicomponent systems, both when it is used for comparative 1-D SDS-PAGE and for 2-D PAGE. The underlying rationale in both cases is that more bands in a gel channel, and more spots over a gel area, can be resolved the longer the migration path, as long as the increase of band width migration distance does not offset the increase of the number of zones which can be resolved from another. This condition of relatively slight increase in zone width with migration distance appears to be fulfilled in 3 cases: SDS-proteins [30], polynucleotides [31] and isoelectric protein zones in EF [32]. The first case is exploited in the 2-D combination of EF and SDS-PAGE on large gel slabs. For nucleotide separations, greatly elongated vertical slab apparatus is available (e.g. Hoefer No. SE 1000).

4.3.3   Design Criteria and Problems

The design criteria for the 1-D vertical gel are not any different from those considered above for tube gel apparatus, except: a) The surface to gel mass ratio is higher, particularly for gels thinner than 3 mm, with consequent advantages in the strength of adherence of the gel to the glass walls. Thus, the various measures for mechanical gel stabilization described for gel tubes assume less importance than in gel tube apparatus. b) The presently available slabs are made of soda glass rather than Pyrex, and the spacers are made of PVC (large slabs) or Plexiglas (disposable small slabs). Since both hydrophilic soda glass and hydrophobic plastics (presumably for different reasons stated in Chapter 5) are not sufficiently adhesive to polyacrylamide gel, this may produce wall separation problems. Coating of the glass with Gelamide 250, agarose or Silane A-174 solves the wall adherence problem on hydrophilic surfaces, while coating of the hydrophobic surface with agarose, or possibly with a non-ionic or amphoteric detergent, would improve their adherence to polyacrylamide. Thus, the use of agarose coating allows one to overcome the adherence problems of soft glass and of plastics simultaneously. c) The apparatus needs to be assembled prior to each experiment, using clamps and spacers. This appears laborious, compared with the rapid insertion of gel tubes or small disposable slabs into the apparatus, and may give rise to leakage problems during polymerization and/or electrophoresis. d) Temperature control in present commercial design is poor, since it is restricted to coolant flow through a few, relatively large bore tubes between the gel slabs. To improve its heat dissipation efficiency, and obtain reproducible R_fs, the entire apparatus needs to be positioned into a thermostated lower buffer reservoir. This can be constructed from a transparent polycarbonate molded box, by placing an electronic cooling device into the container (see Section 4.3.1). However, a MW determination from a standard curve relating MW with R_f in SDS-PAGE by itself does not require reproducible R_fs and therefore an efficient heat dissipation system (Section 15.3).

Vertical 2-D slab apparatus design has the following additional design *problems*: a) It relies to date on the relatively clumsy procedure of having to physically transplant the 1-D gel onto the 2-D gel. A design allowing for "windows" in the spacers through which 1-D electrophoresis could be conducted on the identical composite gel later used, after closing of the "windows" and opening of "windows" at a 90° angle, would appear capable of avoiding such gel transplants but does not exist as yet. b) In present 2-D gel slab design, the gel in the second dimension is really a vertical 1-D gel, which is being loaded by a gel rather than by a liquid sample. However, when as suggested under a), both dimensions of fractionation are run on the same composite gel, a 90° rotation of the slab becomes necessary. It then may be easier to separate upper from lower electrolytes not mechanically, as in present designs, but on the basis of their densities. This can be accomplished e.g. through use of a high density lower buffer containing 20% sucrose separated from a lower density aqueous upper buffer without sucrose by an immiscible heat-conductive layer of intermediate density formed by dibutylphthalate (Eastman No. 1403), phenyl silicone oils or Freon 113-polybutene mixtures of intermediate densities (G. Matthews and A. Chrambach, unpublished data).

4.4 Horizontal Gel Slab Apparatus

Flat bed apparatus has its particular area of excellence in application to mechanically labile gels such as beaded dextran gels (Sephadex), agarose-polyacrylamide "copolymers" (e.g. LKB, Ultrogel) and agarose gels (to the degree that coating of glass surfaces with agarose has not rendered agarose gels compatible with a vertical dimension, i.e. in particular below a concentration of 0.75%). The design requirements for horizontal slabs differ fundamentally from those applied to both gel tubes and slabs in the vertical dimension. The relevant design elements refer to:

- Temperature control
- The gel-air interface and humidity control
- Size of the buffer chambers
- The gel-buffer interface, wicks
- Syphoning
- Sample volume
- Stacking gels
- Compartmentalized apparatus

4.4.1 Temperature Control

Present commercial horizontal gels are *temperature* controlled only through a bottom plate either in contact with circulating coolant or with an electronic Peltier cooling unit. Therefore, a temperature gradient must develop between the cooled and the uncooled side of the slab. This temperature gradient is dependent a) on gel thickness, b) on the thickness and heat transfer characteristics of the insulating layer between coolant plate and gel; and c) on the heat transfer characteristics of the cooling plate. a) With 50 to 250 μ thick gel slabs, heat transfer from any kind of cooling plate, and across any similarly thin layer of insulating plastic (Marine Colloids, Gel-Bond; Serva, Gel-Fix), appears so efficient, even at very high voltages, that for practical purposes the temperature gradient across the gel can be neglected [23],[33],[34]. Thus, the thin-layer technique solves at least the temperature control problem in horizontal apparatus, although at the price of increased artfulness of technique, and, in case of polyacrylamide, surface catalysis problems in the polymerization reaction (Chapter 5). b) In view of the conducting properties of metallic cooling surfaces, insulating plastic or glass layers need to be interposed between gel and cooling plate. To allow for maximal heat transfer, these insulating layers should be as thin as possible and be made of maximally heat-conductive material. Mylar and polyester sheets (e.g. Serva, Gel-Fix sheets of 75 μ thickness) are available for that purpose. Such ultrathin plastic sheets intermediate between gel and coolant plate are less inhibitory to heat transfer than an intermittent glass plate. c) Metallic cooling plates covered with a thin layer of non-conductive plastic (e.g. Pharmacia No. FBE 3000) are preferable heat conductors to glass cooling plates (e.g. BioRad No.1405,1415; LKB No. 2117-301); these are more efficient than plastic cooling plates. A thermoelectrically cooled bottom surface (Hoefer No. HE 950, Isolab No. FR-2100) should be the most efficient since it allows one to compensate for Joule heating at high voltages by correspondingly intensive cooling. d) A horizontal slab apparatus immersed in thermostated liquid has recently been reported [22] which with respect to temperature control should be equivalent to present gel tube apparatus.

4.4.2   Humidity Control

The height of the air-chamber above the gel, and the quality of insulation of that air chamber from the atmosphere, govern the humidity above the gel needed to prevent its partial dehydration, which leads to a gradient of resistivity across the gel. If that height is zero, i.e., if the gel is fully enclosed as a sandwich between two plates as in some paper electrophoresis models, condensation problems appear to arise; it is not clear, though, whether this is also the case when both top and bottom plates are equally cooled. Moreover, sandwich models give rise to design problems in the application of the sample, requiring either "windows" in the top plate or a partially open top plate (e.g. adjacent to stacking gels [29]).

Maintenance of a saturated atmosphere above the gel depends on the height of the air-layer above the gel, the temperature control of the top surface and its insulation. In the design of flat bed immunoelectrophoresis apparatus (e.g. BioRad No.1415), for instance, the maximally tolerable height of that air-layer has been found to be 1 cm for operation at 10–15 °C [17].

4.4.3   The Size of the Buffer Chambers

The size of the buffer chambers must be adequate to maintain essentially constant electrolyte composition (pH, conductance) of the catholyte and anolyte during electrophoresis. This is clearly not the case when the buffer reservoirs are reduced to a filter paper strip saturated with buffer catholytes and anolytes (in contrast to the strong acid and base conventionally used for this purpose in EF), or the positioning of electrodes onto the gel without use of any buffer reservoirs.

4.4.4   Nature of the Gel-Buffer Contact

Direct gel contact with the electrolytes appears preferable to contact mediated by wicks. If wicks are used, perfectly even contact of the wicks across the length of the gel is necessary to provide uniform voltage gradient across all points on the gel. The osmotic properties of the wick have to be the same as those of the gel. The degree of wick overlap with the gel is of importance; 5–8 mm have been suggested as being optimal for the immunoelectrophoresis apparatus discussed in the previous section [17]. The thickness of the wick and the wick material [35] appear crucial.

4.4.5   Syphoning

Syphoning across the gel between catholyte and anolyte necessarily occurs to a degree inversely related to the maintenance of equal buffer levels in the electrolyte chambers, and of horizontal alignment of the apparatus. Thus, the apparatus and gel need to be meticulously aligned on a levelling plate.

4.4.6   Sample Volume

Sample volume is far more restricted than in the vertical gel techniques. A suitable technique with agarose gels giving very sharp starting zones for sample volumes not exceeding 10 μL consists of the loading of the protein sample into slits in the gel which have been previously desiccated by insertion of a filter paper strip and buffer withdrawal by capillary action to a height of 1 – 1.5 cm. The slits contract after withdrawal of the paper strip, thus providing a very thin starting zone and correspondingly good resolution [36]. A more conventional manner of sample application uses paper strips saturated with sample solution, with obvious need to consider adsorption, strip material, dehydration etc. An example for a successful technique is provided by [35].

4.4.7 Stacking Gels

When the gel is poured horizontally, stacking gels with parallel surfaces to those of the resolving gel can only be formed after excision of a gel slice. The excised space is then filled with stacking gel. The gel excision raises the problem of mechanical stress on the gel, leading to wall separation in application to relatively elastic polyacrylamide [37]. Agarose gels lend themselves better to that technique, but have limitations with regard to pore size and to stacking which will be discussed in Chapter 6. The problem does not arise when horizontal gels are formed vertically [38].

4.4.8 Compartmentalized Apparatus

One of the key advantages of tube apparatus in Quantitative PAGE is the possibility to conduct electrophoresis at various gel concentrations simultaneously. This has recently become a possibility for agarose gel electrophoresis as well, through the design of horizontal compartmentalized apparatus [22],[39],[40]. However, it should be realized that in the horizontal partitioned apparatus described in [22] the assumption is made that the voltage drop through the channels is the same, independently of gel concentration. While this is true as a first approximation for the range of agarose concentrations up to 1 or 2%, it is certainly not true for polyacrylamide [26] and presumably not at higher agarose concentrations. For the same reasons, "submerged" conventional agarose gels give rise to voltage drops across the gel (and thus protein mobilities) not significantly different from those with open top surfaces, while that would not be expected to be true at higher agarose concentrations or with polyacrylamide. The present commercial horizontal gel apparatus does not allow one to excise channels at different times of electrophoresis without mechanically labilizing the remaining gel. This constitutes a problem in electrofocusing (see Section 12.1).

In summary, the design of horizontal slab apparatus raises a number of special problems distinct from those of either tube or vertical slab apparatus, which make its application vastly more artful and in need of meticulous control of conditions than that of other apparatus types. Yet, there are two areas where its use seems indispensable. First, when very open pore sizes are required, as in many preparative problems in the interest of recovery [34], or in application to very large molecules or particles [40], one has no choice except to deal with these design problems of horizontal gel slab apparatus. Secondly, when one is faced with very large numbers of daily analyses, usually in the hundreds, as in clinical analysis or studies on large populations [29]. In particular, agarose is a suitable matrix for such applications in view of the extreme simplicity of its gelation procedure [36].

4.5 Multichamber Gel Tube Electrophoresis Apparatus

The function of multichamber tube apparatus is to make it more convenient to carry out gel electrophoresis in different buffer systems simultaneously. Such analysis is routinely needed in the optimization of the pH of PAGE by electrophoresis on stacking gels made in different buffer systems (Chapter 7). Another use of this apparatus is in detergent optimization, allowing one to test simultaneously stacking gels made at different detergent concentrations [41].

Multichamber gel tube electrophoresis apparatus has been reported both as an adaptation of an commercially available Plexiglas tube apparatus [41] (Fig.6A) and as an all-Pyrex construction [42]. The latter model although available (R & D Scientific Glass Co.) is relatively expensive, in view of its mode of construction by Pyrex fusion (Fig.6B). The former apparatus is not as yet commercially available, but can be relatively easily made in the laboratory workshop from gel tube apparatus Model IV (Hoefer) by cementing Plexiglas cylinders over each of the holes in the upper buffer reservoir which hold the rubber grommets and gel tubes, as well as into the lower buffer reservoir in alignment with each gel tube (Fig.6). The only further modification needed is a re-wiring of the electrode wires in the upper and lower buffer reservoirs. The easiest mode is to pass the electrodes along the circumference of the bottom plates of both the upper and lower reservoirs before cementing in the Plexig-

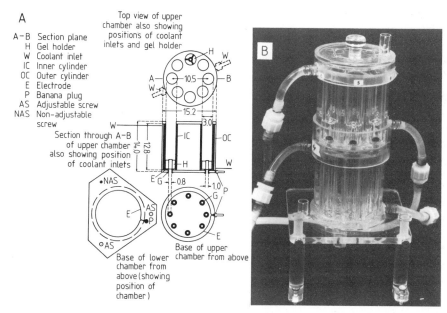

Figure 6. Multichamber gel tube apparatus. (A) Mode of construction in the workshop, by modification of tube apparatus Model IV (Hoefer) [41]. (B) Pyrex construction (R & D). Individual upper and lower buffer reservoirs for each gel tube are separated, to allow for the simultaneous analysis of stacking gels differing in buffer system.

las cylinders. Care must be taken, of course, that the wires pass through each of the chambers. The main defects of this design are that the cementing of the Plexiglas cylinders may loosen, leading to leakages after some use. Secondly, the electrode wire connections between the bottom plates of the reservoirs and the corresponding banana plugs on the outside of the apparatus need to be safely insulated, and in the reported design [41] even pass through the water jacket of the lower reservoir. The Pyrex model avoids these problems, but presents others: The present model positions the lower electrodes symmetrically at the center of the chambers, giving rise to gas accumulation under the gel tubes during electrophoresis. A desirable design feature of any apparatus for use with stacking gels is hydrostatic equilibration, i.e. equal buffer levels of upper and lower buffer reservoirs preventing gravitational stress on the relatively labile open-pore gels (see Fig.5). Neither of the above-described multichamber apparatus models incorporates that desirable feature. The Pyrex model has the further shortcoming of being too tall, requiring gel tubes of 25 cm length (shorter tubes may be used, but then sample application is difficult). Since the gel length of a stacking gel of 1 to 2 times the tube length occupied by the sample is sufficient (see below), and samples usually do not exceed 1 mL volume, a design for 5 cm tube length would certainly suffice.

4.6   Pore Gradient Gel Electrophoresis Apparatus

Pore gradient gel electrophoresis is of importance in multicomponent analysis by providing for each component pair the optimally resolving pore size (Chapter 11) for at least one point along the migration path, and for providing useful migration distances for molecules of widely divergent molecular weights. A further advantage in the use of pore gradients consists of the fact that bands with long migration distances remain relatively sharp; resolution is therefore increased [43].

Another application of pore gradients has been the estimation of molecular weights on the basis of the migration distance at which migration velocity of a particular band is sufficiently decreased that it appears to have reached a "dead stop". In theory, such an arrest of bands is impossible since a pore size distribution exists at each gel concentration which allows for passage of at least some large molecules through the relatively small proportion of open pores [44]. In practice, however, the "dead stop" is a useful approximation which empirically provides a good correlation with molecular weights [45].

In all the above applications of pore gradients, electrophoresis proceeded along the axis of the pore gradient; we might designate this case as "longitudinal pore gradient electrophoresis". Another important application is that in which electrophoresis is carried out at a right angle to the axis of the pore gradient [46],[27]. This case is designated as "transverse pore gradient gel electrophoresis". Its usefulness is to test separation not at one, but simultaneously at many gel concentrations, including at one point the optimal gel concentration for separation between any specific pair of species. Thus, the curvilinear gel pattern characteristic of each component in transverse pore gradient gel electrophoresis represents an experimental, non-logarithmically plotted, Ferguson plot. It is an essential tool for the assignment of bands in multicomponent analysis by PAGE to particular Ferguson plots [47] (Chapter 10). Yet another type of pore gradient electrophoresis is a transverse pore gradient gel used in the second dimension and loaded with a cylindrical longitudinal pore gradient gel (Fig.1 of [27]). This arrangement allows for a separation in the 2nd dimension of each separated component at a gel concentration approximating the optimal gel concentration for its separation from the neighboring zones. The apparatus described below for use in transverse pore gradient electrophoresis (Fig. 2 of Appendix 4) can be applied to this type of 2-D electrophoresis.

4.6.1 Longitudinal Pore Gradient Electrophoresis, Gradient Maker

Longitudinal pore gradient electrophoresis does not require any special apparatus other than a gradient *mixer*. Tube or any type of vertically oriented slab gel electrophoresis apparatus can be selected by the criteria which have been elaborated above. The design of a gradient mixer applicable to polymerization mixtures at any pH and temperature selected for gel electrophoresis has to take the polymerization reaction (Chapter 5) into account which gives rise to a crosslinked gel: a) the need to regulate the temperature of polymerization at the temperature of electrophoresis; b) the need to achieve a high degree of conversion of monomers to polymer as well as a high average chainlength. These conditions define a reproducible time course of polymerization which is particularly required for the polymerization of gradient gels in order to achieve equally high conversion of monomers to polymer, and equally high average chainlength at both extremes of the pore gradient. Necessarily, polymerization conditions also have to prevent premature polymerization while the pore gradient is being formed. A jacketed 2-chamber gradient mixer for polymerization mixtures has been designed in accordance with the need to control the rate of the polymerization reaction [48] (Fig.1 of Appendix 4) and a procedure for its use has been described (Chapter V, B8 of [17]). The solution to the problem of an airlock between the chambers which forms when they are being filled is to use a 3-way stopcock connected to a hollow cylinder. With the stopcock turned so as to block the line between chambers, solution is drawn by syringe into one half of the line and into the cylinder; the contents of the cylinder are returned to the chamber. Then the procedure is repeated with the second chamber. The gradient maker is used in conjunction with a proportioning pump (Technicon). Flow lines outside the pump may be jacketed and protected from light (the latter to prevent premature photopolymerization). A commercially available water jacketed model, if protected from light by wrapping in aluminum foil, appears also applicable (Hoefer No.XPO 77; BioRad No.230). The discussion of the various design elements of such a mixer will also equip the reader with the basic concepts needed to modify the construction and/or improvise the use of one of the many simple 2-chamber gradient makers on the market (e.g. Isolab No.PGM-1 to PGM-20 and GM-1; Hoefer No. SG 101S to 105S; MRA No.116).

It should also be noted that certain pore gradient gels are commercially available in prefabricated form (Isolab No. GC-1 to GC-7 ISA-1, ISB-1). The problems related to prefabricated gels in general will be discussed in Chapter 5.

4.6.1.1   Temperature control

Temperature control of the polymerization mixture is needed to dissipate the heat of polymerization which is considerable at high gel concentrations, and to prevent thermal contraction/expansion from introducing gradient inhomogeneity along the cross-section of the gel, and to prevent wall separation. Placing the entire apparatus into the cold room, and immersing both gradient maker and slab apparatus in a water bath, preferably with magnetic stirring, should be able to substitute for the temperature control by means of water jackets to a large degree. A more precise and costlier way to thermostat the gradient maker would be by electronic cold finger and temperature regulator, in the manner described in Section 4.3.3.

4.6.1.2   Polymerization catalysts

To obtain a uniformly well-polymerized gel polymerization at all gel concentrations should proceed within 10 min (Chapter 5). Thus, polymerization catalysts and inhibitors in each of the 2 chambers of the mixer should be selected so as to provide that particular polymerization rate. To largely eliminate oxygen inhibition, saturation of both chambers with Argon is advisable (see Section 4.7). If the pH is acid or neutral, and photopolymerization is consequently required (Chapter 5), the gradient mixer, flow lines, and apparatus should be covered by a black cloth or by Al foil, while the polymerization mixture flows into the slab apparatus.

4.6.2   Transverse Pore Gradient Electrophoresis

Transverse pore gradient electrophoresis requires, in addition to the gradient mixer described above, a special form of slab apparatus. A prototype of such apparatus has been described (Fig.2 of Appendix 4; Fig.17 of [17]) but is not sufficiently simple in design or commercially available. In all applications of transverse pore gradients to date such apparatus has been improvised by removing the spacers after polymerization, and re-application of spacers at a right angle, with insulation through greasing. During electrophoresis, such an improvised construction may obviously lead to electrical leakage.

4.7   Apparatus for the Deaeration of the Polymerization Mixture

Since oxygen is an inhibitor of the polymerization reaction of polyacrylamide (Chapter 5), it must be removed from the polymerization mixture a) to provide a polymer in those cases where dissolved air in the polymerization mixture would prevent a sufficient degree of polymerization or delay the onset of polymerization unduly; b) to allow one to polymerize reproducibly, i.e. at a reproducible rate. The latter is essential for resolving gels; all of Quantitative PAGE rests on the assumption that pore sizes can be formed reproducibly.

For the purposes of stacking gels, polymerization must not necessarily proceed quantitatively (in contrast to the resolving gels); all that's required is a sufficient degree of mechanical stability to hold the gel firmly in the tube. The duration of polymerization is also less critical than with resolving gels. For convenience, however, it should not exceed 0.5 h. Thus, deaeration should be applied to the polymerization mixtures of stacking gels mainly to reduce the necessary concentrations of polymerization catalysts (see below), to overcome inhibition by protons at acidic pHs (and to a lesser degree at neutrality) or inhibition by buffers which can act as free radical captors (e.g. those with aromatic groups).

For all of these purposes, deaeration of the polymerization mixtures by timed evacuation with an oilpump or an efficient water aspirator (capable of providing

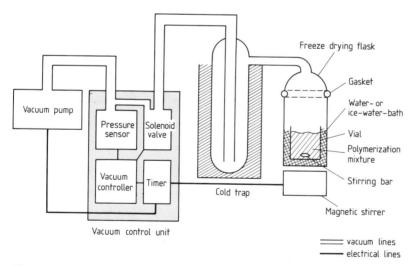

Figure 7. Apparatus for the controlled deaeration of polymerization mixtures [49]. Lyophilization flask (Thomas No.5119-B22); Vacuum controller with gauge tube No.DV4R (Teledyne Hastings-Raydist Co., POB 1275, Hampton VA 23661), solenoid valve (Skinner Electric valve No.V52DB2100 110 VAC Airline Hydraulics Co., 107 Beaver Court, Cockeysville MD 21030), solid state relay (No.601-1402 Teledyne-Relays, 3155 W El Segundo Blvd., Hawthorne CA 90250). The solid state relay operates on a TTL logic level. With 3-32 V input control voltage, it gives a 10 A output. Timer; 3-way stopcock; Dewar flask electronically cooled (Neslab CryoCool); Oilpump.

15–20 mm Hg) is sufficient. It is only in view of the more stringent requirements for deaeration in Quantitative PAGE (Chapters 10,11), that a more efficient and reproducible mode of deaeration needs to be set up. Once set up, it is most conveniently also used for stacking gels.

Deaeration can be controlled by temperature (at uncontrolled pressure) or by pressure (at uncontrolled temperature). However, the former approach is experimentally not easy, since the polymerization mixture rapidly cools upon evacuation. The procedurally simplest but most expensive mode of deaeration without rigid temperature control is by an electronically controlled oilpump [49]. A less costly alternative is deaeration by timed and regulated flow of Argon through the polymerization mixture. Fig.7 depicts schematically the apparatus for controlled deaeration by oilpump. The polymerization mixture (minus TEMED) is placed into a 600 mL lyophilization flask. Activation of the oilpump simultaneously starts operation of the electronic vacuum controller (set to 10 mm Hg at 0 °C, 20 mm at 25 °C), the magnetic stirrer under the polymerization mixture and the timer (set to 5 min).

For deaeration with Argon, only a flow controller (e.g., Cole-Parmer, Chicago IL, No. 3235-11) set at 20 mL/min is needed. The gas is bubbled through the thermally controlled polymerization mixture at the controlled rate for a fixed time (e.g. 5 min), thus displacing the dissolved air. Argon, being heavier than air, settles above the polymerization mixture and thus protects it from oxygen during the dispensing of the polymerization mixture into the apparatus.

Both deaeration procedures give rise to bubbling if the polymerization mixture contains protein or detergent. It is therefore advisable to add protein [50] or detergent *after* deaeration whenever possible, not before.

4.8 Photoinitiation Apparatus

Polymerization at acid and neutral pHs requires photoinitiation (Chapter 5). No photochemical studies exist to date which would provide a solid basis for the design of light sources optimized with regard to both light intensity and wavelength in application to crosslinked polyacrylamide gel. The conventional photoinitiator in PAGE, riboflavin, absorbs at 445 and 375 nm [51], suggesting the construction of light sources in that range. Efficiency of riboflavin initiated photopolymerization of linear polyacrylamide is large as compared to any other controllable free radical initiated polymerization [51]. This suggests, that relatively low intensities suffice. In practice, symmetrical (concentric) illumination of the gel by an array of 6 or 9 twenty-watt daylight fluorescent tubes appears adequate (Fig.3 of Appendix 4). Suitable commercial apparatus has been available (Buchler No.3-1712 "preparative" illuminators) but appears discontinued at this time. An available commercial unit that lends itself to some gel apparatus geometries consists of circular fluorescent light units stacked one upon the other (Hoefer No.EF 320). Another commercial model

(Buchler No.3-1778) comprises light sources of considerably less height and a total of 75 watts. It should be an adequate illuminator for analytical apparatus which does not exceed its height. An inexpensive, home-made set of concentric illuminators may be constructed by mounting the requisite number of 20 Watt tubes into the 2 halves of a vertically sliced plastic waste bucket.

4.9   Apparatus for Overlayering and Sample Application

Overlayering of the polymerization mixtures is needed to avoid curved gel surfaces (a meniscus) and to diminish exposure of the gel surface to atmospheric oxygen which is inhibitory to free radical polymerization. Water or gel buffer are usually used for overlayering. Where it is essential to prevent dilution of the surface of the polymerization mixture, sec-butylalcohol can be applied; it is an inhibitor of polymerization like other aliphatic alcohols and therefore provides, immiscible at its interface with the aqueous polymerization mixture, a very sharp gel surface if initiator concentrations are sufficiently high. It has the disadvantage of not being quantitatively removable from the gel surface, with consequent possibility of protein inactivation at the surface.

4.9.1   Overlayering Devices

Overlayering devices. Any device capable of applying the overlayering solution to the surface of the polymerization mixture gently enough to prevent its convection ("bombing") is applicable. a) The simplest such devices are a lubricated (Lubriseal, Thomas No.8690-B10) 1 mL glass *syringe* with 23 gauge hypodermic needle (preferably with blunted tip) of 2.5 to 3 inch length (e.g. 23G × 2.75" spinal needle, Terumo Corp., Tokyo, Japan, No. SN 2370). b) An even more gentle way of overlayering employs a 500 μL Hamilton syringe (gas tight, No.81217). c) A syphoning device, using capillary tubing can also be used; but the curved plastic tube is not as easily positioned inside the gel tube or slab than a rigid needle.

4.9.2   Sample Application Devices

For sample application, either a gas-tight Hamilton syringe of appropriate volume is being used (Nos. 80900, 81000, 81217) or a Kirk micropipet of suitable volume (Thomas No.7660-C42 to -C84) in conjunction with a micropipet syringe control (Microchemical Specialties, Berkeley CA 94707, No.MB-1.0). Constricted (Lang-Levy) micropipets do not drain evenly and are therefore not suitable. Mouth control of the micropipets also is not as useful in sample application as syringe control since it does not provide as even a pressure and makes it more difficult to observe the gel surface while overlayering. Eppendorf type pipets have the disadvantage in sample application into gel tubes that their tips are too wide and too short to allow one to dispense the sample onto the gel surface.

4.10   Power Supply

4.10.1   Wattage Control

Since resolution increases with field strength, an attempt should always be made to conduct electrophoresis at the maximal voltage compatible with the capability of the apparatus to dissipate Joule heat. Therefore, conceptually it would be best to regulate wattage at the maximal level that allows one to effectively dissipate the Joule heat. In practice, however, this rationale is not a useful guide in application to one or a few cylindrical gels (with relatively small gel surface area), because present-day power control at sufficiently low wattages (below 1 watt) is not available. In application to PAGE in discontinuous buffer systems, wattage control furthermore is ill-defined, since it refers to the average voltage across a gel consisting of 2 gel phases, in addition to the upper and lower buffers, across all of which the voltages differ. Finally, in application to EF on gel slabs, it is usually not possible to maintain power control throughout the run, since with decreasing current the voltage requirement exceeds the voltage capacity of the supplies, with the consequence that at some usually unobserved time wattage control is supplanted by voltage control.

4.10.2   Current Control

Current control at 4 to 8 mA/cm^2 of gel at 0 °C, twice that much at 25 °C, is convenient in the practice of gel electrophoresis, and in electrofocusing during the first transient state, i.e. while the carrier ampholytes migrate to their isolelectric positions, thus setting up the pH gradient. This level of current does not give rise to excessive Joule heat with consequent thermal denaturation of the protein (except in preparative moving boundary electrophoresis (Chapter 13) or in application to heat-sensitive species, e.g. [52][105]), even at the end of an electrophoretic run when the voltage is highest, provided that apparatus geometries and heat dissipation characteristics correspond to those typical for the tube apparatus Model I described above. However, current regulation at those safe but arbitrary levels fails to provide, at anyone time of electrophoresis, the maximum tolerable voltage and thus the best possible resolution.

4.10.3 Voltage Control

Voltage control is the preferred mode during the steady-state of EF and MBE. A voltage drop of at least 20 V/vertical cm of gel is tolerable in EF in the above-described gel tube apparatus. The value is much higher, of the order of 100 to 500 V/cm of gel, when the Joule heat is dissipated through a gel not thicker than 0.25 mm and an even thinner film of Gel-Bond by means of Peltier effect cooling [23]. In preparative MBE [53], due to the enormous protein concentrations within the stack, the likelihood of protein denaturation through excessive Joule heating is much augmented, and voltages have to be reduced to as little as 2 V/cm.

Thus, in application of the techniques described in this book, current and voltage regulated power supplies with a current regulation range between 1 and 100 mA, and a voltage regulation range between 100 and 2000 V are perfectly adequate. The use of unregulated power supplies should be avoided in the interest of reproducibility of electrophoretic conditions.

Correspondingly, voltage control of at least 5000 V are desirable in application to Peltier effect cooled thin layers of gel described above [23]. Such regulated power supplies are presently not available.

4.10.4 Pulsed Power

A potentially important mode of power supply in gel electrophoresis is "pulsed power" (Ortec No.4120 Pulsed P/S, MV 159). Although "pulsing" at identical wattages cannot provide a saving in Joule heat as has been sometimes erroneously implied, it does have the potential advantage that it allows one to use apparatus with any heat dissipation characteristics at the maximal wattage by allowing one to compensate poor heat dissipation by a correspondingly longer duration of cooling periods provided by the off-periods between pulses. Conceivably, therefore, one may be able to construct pulsed power supplies operating at the maximal voltage, compatible with a desired value of temperature inside the gel, by connecting the power supply to a thermistor implanted in the gel and allowing for control of voltage by the thermistor. This should allow for the best possible resolution in any apparatus, with electrophoresis times higher when the heat dissipation characteristics of the apparatus are poorer, and *vice versa*.

4.11   Apparatus for Gel Evaluation

4.11.1   Tools for Removing Gels from Tubes or Slabs

Polyacrylamide gels are separated from glass surfaces in either one of 2 ways: By reaming with water, or by smashing of the glass tube and peeling off the splinters under water. The latter method is applied to hard gels in excess of 12 to 15 %T, 2 to 5 %C_{Bis} and requires solely a heavy hammer and stone surface. An alternative suitable device, consisting of an aluminum block with central groove, holding the gel tube, is available (BioRad "Gel-Eliminator"). The reaming method is applicable in all other cases and requires a 23 gauge blunt 2 to 3 inch long hypodermic needle (e.g. 23G × 2.75 inch, No. SN 2370 spinal needle, Terumo Corp., Tokyo, Japan) through which water is passed continuously and evenly either through a water line connected to the faucet, or to a syphon. With the tube held vertically, the needle is progressively inserted along the inner tube wall and the tube is rotated (firm gels) or the needle is withdrawn and reinserted at 45 degree intervals (weak gels). Then this procedure is repeated from the opposite gel end. Reaming from alternate sides is repeated until the gel slides on the bench surface. The intensity of the water stream emanating from the needle is increased for firmer gels and decreased for weaker gels. Using a needle valve in the line is convenient in controlling the stream. Gels are easily damaged when one tries to lift them by forceps. Manual lifting is also inadvisable, since fingerprints are visible on the stained gels. The best procedure of lifting gels is to slide the gels by spatula along the bench surface and onto a curved scoopula (Thomas No.9006-L2). If necessary, one should wet the bench surface.

Agarose gels are removed from gel tubes by inserting the reaming-needle vertically along the inner tube walls at a number of points along the circumference of the gel, without rotating either tube or needle. Then, a solid glass rod, closely fitting the ID of the tube, is inserted as a piston from one end of the gel and the gel is gently pushed out of the tube onto the bench surface.

Sephadex gels are removed from tubes after freezing of the gels in liquid nitrogen, and peeling off the cracked tube on the bench surface. If large glass pieces adhere, or the tube should not crack, the polyacrylamide plugs are cut off, and the gels gently pushed with a glass rod as in the case of agarose [54].

4.11.2   Apparatus for Staining

Fixation and staining of gels depends on the final concentration of fixative and stain after immersion of the gel. Therefore, to be able to neglect the gel volume of water, it is necessary to use a large excess of fixative; usually a 40-fold excess is applied, requiring a 40 to 60 mL vial or cylinder per gel. Snap-on polypropylene bottle tops are preferred because they are resistant to fixatives. Since most tube apparatus contains 10 to 12 gel tubes, a bottle rack for that number of bottles is convenient. Correspondingly large dishes are used to fix gel slabs.

If gels contain detergents which interfere with fixation and staining, they must be diffused from the gel in a fixative which precipitates the proteins under investigation quantitatively. Diffusion destainers are available for that purpose (Hoefer No.SE 530; BioRad No.165-0910), or more effective ones can be constructed from perforated, acid resistant tubes [55] (Fig.4 of Appendix 4). The same apparatus can also be used for destaining of stained gel background as is inevitably obtained if detergents are not removed prior to staining. Electrophoretic removal of either charged detergents or stained background employs either 10 mm ID tubes constricted at one end in an ordinary tube gel apparatus, or an apparatus for transverse electrophoresis through cylindrical gels (Miles No.43-535-1). But this technique suffers from incompatibility with reliable fixatives such as TCA, and risks that zones continue to migrate at least prior to saturation of the gel with fixative.

Photography of cylindrical gels is best carried out in clear glass tubes containing fixative. Thus, clear, unmarked glass tubes with wide mouths and Teflon or polypropylene lined screw caps are needed for gel photography (Section 4.11.7) and storage.

4.11.3   Transverse Gel Slicers

Four types of gel slicer have their particular areas of application: a) For all solid gels, i.e., 4 to 40 %T, 2 to 5 %C_{Bis}, the Hoefer slicer with electronic vibrator (Hoefer No.SL-280; BioRad No.165-2500) is useful but does not guarantee slice order or slice recovery as well as model b). It is available for gel diameters of 3, 5 and 14 mm but can be modified for application to other gel dimensions. Application to 5 and 6 %T, 15 %C_{DATD} gels is marginal and may require some apparatus modifications, particularly with wide diameter gels. Since these gels are elastic, they cannot be sliced unless the wire passes across the entire cross-section. The groove in which the gel is seated should be fitted with a floor of 2% agarose; the agarose gel allows for passage of the wires through the entire cross-section of the gel [56]. Alternatively, a grid may be inserted into the slicer which allows for passage of the wire blades through the entire cross-section of the gel cylinder [57]. In application to these and other soft gels (4 %T, 2 to 5 %C_{Bis} or less) the vibrator must be decelerated by rheostat. Large diameter gels particularly are easily disrupted unless so decelerated. b) In application to the same range of solid gels defined under a) a manual slicer with a concentric iris diaphragm cutting edge represents a relatively economical alternative [58] (Fig.5 of Appendix 4). This model needs to be built in the user's workshop, but construction consists mainly of an assembly of commercially available parts (see Fig.18 of [17]). c) For large diameter (e.g. 18 mm) 5 to 6 %T, 15 %C_{DATD} gels, a manually operated wire slicer [59] (Fig.6 of Appendix 4; Fig.3 of [53]) seems best suited. This slicer type requires a moderate degree of hardening of the gel through freezing, the degree of hardening depending on gel concentration and crosslinking. d) Granular gels, are presently sliced manually after freezing [54]; a razor blade guillotine device (Mickle Labs., Gemshall, Surrey, UK; also produced by Joyce-Loebl) appears potentially applicable.

4.11.4   Longitudinal Gel Slicers

Longitudinal gel slices are of importance in autoradiography, or in application of gels to multiple staining procedures, or in provision of a 1-D gel sample in 2-D gel electrophoresis. The most useful type of longitudinal gel slicer appears to be one with a heavy, metallic stand for operation on the bench, and with unidirectional movement of the cutting wires and little tolerance for their lateral motion [60] (Fig.7 of Appendix 4; Fig. 11 of [17]) which needs to be constructed in the workshop for gel cylinders of the desired diameters. A commercially available version of this slicer (Miles No. 43-560-1) appears less satisfactory in performance presumably because of the absence of shims or equivalent ways to allow for passage of the cutting wires through the slots with minimal resistance, while the gel is encased in the block of the slicer under uniform pressure (produced by an even tightening of set screws in the original model).

4.11.5 R_f Measuring Devices

Bands in gel electrophoresis are characterized by their absolute or relative migration distance. Migration distances can be estimated within a fraction of a mm on cylindrical gels themselves, using a ruler and calipers under a magnifying glass, or by such measurements carried out on gel photographs. A more exact and less fatiguing way to measure migration distances is electronic, by use of a sliding wire potentiometer [61] (Fig.8 of Appendix 4). A similar model is available as a commercial prototype (Hoefer). An even more sophisticated model providing a magnifying screen is also produced commercially (Brinkmann, Desaga "Multiskop" No. 141202).

4.11.6   Gel Drying and Storage Equipment

Drying of longitudinal gel slices of cylindrical gels or of gel slabs is best carried out on a commercial device (Hoefer No.SE540; BioRad No.165-0920; Miles No.43-526-1) allowing for application of vacuum and, at low concentrations, heat. The addition of agarose (Indubiose, IBF, Gennevilliers, France) or glycerol [194] to polyacrylamide gels greatly improves its drying characteristics. The same is true for polyacrylamide gels crosslinked by vinyl-agarose [62].

Gel storage requires a sealed, acid resistant and dark container. For cylindrical gels, a dark sample storage box for 13 × 100 tubes divided into regular arrays to alleviate tube identification and retrieval is the most useful. Storage tubes should be wide-mouthed and screw-capped. Screw caps should have an acid-resistant lining. Tube storage boxes (Cargille No. SS100) are convenient for purposes of organizing the gel data and data retrieval.

4.11.7   Photographic Equipment

An inexpensive Polaroid camera adjusted to constant focal length by suitable lens arrangement and combined with a gel tube or slab holder and suitable illumination has not been commercially assembled as a unified design to date. The closest approximations are a Polaroid camera assembly (used with No.107 film) by Miles (No.43-523-1) and a prototype by Hoefer (No.CS50). But it is dubious that either one provides the desirable degree of contrast and of photographic size for each gel. An MP-3 Polaroid camera (No.52 film) is a far more expensive approach to gel photography, but it also fails to provide adequate contrast for publication purposes. To obtain gel photographs which reflect all visible detail one must use this or any other camera with a superior lens (Tessar 1:45, F = 150 mm, No. 18453, or equivalent, in conjunction with Ektapan 100 ASA film, Kodak No. 1689777, or equivalent). A suitable illuminator for transmitted light is an X-ray illumination box. Yellow filters appear to enhance contrast of protein bands stained with Coomassie G- or R-250.

To obtain photographs of unstained protein precipitate bands in EF, dark field illumination (Miles No. 42-100) and Polaroid film 57 (MP-3 camera) can be used.

4.11.8   Zone Scanning Apparatus

Densitometry of stained bands claims to yield quantitation but only provides it in rare cases. Quantitation of zones by densitometry can only be obtained if all protein zones under consideration have very similar dye-binding characteristics (molar extinction coefficients or color values). If that is the case, it depends on the constancy of densitometric area under each band as a function of migration distance. If that's given, a standard curve of area *vs* protein loads is needed. To assure inter-experiment reproducibility of quantitation, the most stringent type of control over the staining reaction is required, unless a covalently linked chromophore is used for densitometry, e.g. UV-absorption.

Many useful "gel scanners" are commercially available. For the visible range, a scanner with Laser optics (Biomed Nos. SL-TRFF, SL-504, SL-TRF; LKB No.SLSD) appears popular notwithstanding the fact that Beer's law is not obeyed in such optics. A highly satisfactory but expensive gel scanner for visible or UV light is one produced by Zeiss (Chromatogram spectrophotometer, Druckschrift 50-657 K/D). No present model provides quartz tubes or tube holders for the densitometry of cylindrical gels contained in fixative.

4.11.9   Autoradiography Apparatus

Gel autoradiography requires a cassette (Kodak 120 7299), intensifying screen (DuPont Lightning Plus-Screen, No.54303, 8 × 10 inch) for β-emitters, X-ray film (Kodak X-Omat AR, 8 × 10 inch), lighttight exposure holder (Kodak No.149 2776) and equipment for the development of X-ray film.

4.11.10   Contact pH Electrodes

For the determination of pH gradients in EF, and for pH analysis across the stack in MBE, gels need to be analyzed for pH. This can be done by transverse gel slicing, placing each slice into a dilute salt solution (e.g. 0.5 mL 0.025 M KCl; Na^+ should be avoided in view of the Na^+ error), and manual pH measurement on each. In the practice of EF, manual sectioning of the gel with a single-edged razor blade to give 16 slices appears frequently adequate.

A more sophisticated way to measure pH across gels is by means of a contact electrode, with measurements at selected intervals (e.g. 5 mm). The mechanically most stable setup consists of metal frame constructed to allow for vertical and horizontal movement of the electrode (Brinkmann, Desaga No. 122060 with contact electrode No. 122052 or Ingold No.LOT-403-30-M8). This setup allows one to keep the electrode at one position on the gel until a constant pH is reached. It is therefore largely independent of the age of the electrode.

A fully automated instrument (Hoefer No. GS 127 with Ingold electrode No.6122) which provides a recording of the pH profile has been developed [63] but has the disadvantage that upon aging of the electrode, and decrease of its response time, longer and longer passage times of the electrode along the gel are required, and that no assurance exists at any time that the continuously moving electrode has stopped drifting at any one pH value.

4.12   Computer Programs and Output

Quantitative PAGE requires two types of computer program or output.

4.12.1   The Extensive Buffer Systems Output

The program devised by T. M. Jovin provides buffer systems capable of MBE at any pH. The program and a representative model output designed for proteins in microfiche form are available from the US Department of Commerce, National Technical Information Service at nominal price, as detailed by Table 1 of Appendix 2. Fig.1 of Appendix 2 [13] lists the buffer constituents on which the "Extensive Buffer Systems Output" is based, together with their pK s and ionic mobilities at 0 and 25 °C. Fig. 2 of Appendix 2 shows a representative page of the catalog of the 4267 buffer systems of the "Extensive Buffer Systems Output" [14]. Figs.3 to 5 of Appendix 2 depict the computer output of a representative buffer system. Fig.6 of Appendix shows a representative input page for running the Jovin program for the purpose of obtaining an explicit output of one of the subsystems of Fig.5 of the Appendix.

4.12.2   The PAGE-PACK

The PAGE-PACK, devised by D. Rodbard, computes Ferguson plots, molecular weights, and valences, and optimum gel concentrations for resolution of any set of two components on the basis of relative mobility (R_f) data. It is available from the Biomedical Computing Technology Information Center, Room 1302, Vanderbilt Medical Center, Nashville, Tennessee 37232, No. MED-34 PAGE-PACK, as a magnetic tape with documentation; programs are written in FORTRAN. Fig.8 lists the components of the PAGE-PACK and their arrangement in the form in which they are used at the NIH in conjunction with an IBM 2741, 7000 or 8188 terminal located in the laboratory and connected on line via time-sharing to an IBM 370 computer. Programs are not applicable to other computers without minor modifications. Program names correspond to those given in the instructions. Figs. 9–19 illustrate representative input and output formats of the key programs of the PAGE-PACK.

PAGE-PACK   David Rodbard

Part	Information	Data file (card)	User's program	Source program	Load module	Output
I	PAGEPACK	DATA01	RFT1	RFT1.SOURCE	FERGUSON	F-plot K_R, Y_0
II	and	DATA02 RADII4	PLOTRUN	RADKR1	MWKRPLOT	$\bar{R}$, MW
III	PAGE.PROG	DATA04 RMDATA RADII4	CHARGE			M_0, V
IV		DATA06	TOPT	TOPT.SOURCE	GELOPT	T_{max} T_{opt}
V		DATA03	GIANTRUN	GIANT.SOURCE	GIANTMOD	MW

Figure 8. Components of the PAGE-PACK (D. Rodbard). Each of the components is stored in an IBM-370 computer and accessed via a remote terminal in the laboratory and use of time-sharing.

Input File DATA01

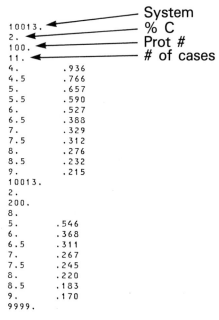

Figure 9. Representative PAGE-PACK input file DATA01 for program RFT1. The left column denotes %T, the right column R_f; 9999 denotes the affirmative command to provide a graphic Ferguson plot.

Program RFT1

Output 1

```
THURSDAY            MARCH   8, 1984

SYSTEM 10013.000    CROSSLINKING    2.00    PROTEIN  100.00

T        RF       PREDICTED        LOG(RF)    WORKING     WEIGHT
                     RF                        LOG(RF)

   4.00   0.9360   0.91285        -0.02872   -0.02859    0.23809
   4.50   0.7660   0.78254        -0.11577   -0.11567    0.21584
   5.00   0.6570   0.67083        -0.18243   -0.18234    0.19149
   5.50   0.5900   0.57507        -0.22915   -0.22900    0.16601
   6.00   0.5270   0.49298        -0.27819   -0.27720    0.14055
   6.50   0.3880   0.42260        -0.41117   -0.40963    0.11629
   7.00   0.3290   0.36228        -0.48280   -0.48085    0.09417
   7.50   0.3120   0.31056        -0.50585   -0.50584    0.07480
   8.00   0.2760   0.26623        -0.55909   -0.55881    0.05845
   8.50   0.2320   0.22822        -0.63451   -0.63445    0.04505
   9.00   0.2150   0.19564        -0.66756   -0.66557    0.03433


UNWEIGHTED REGRESSION
MEAN X=  6.50   MEAN Y=  -0.3723    ANTILOG(MEAN Y)=  0.424331
Y INTERCEPT A =     0.47354E+00    STD.DEV. A =   0.032823
ANTILOG A =    2.97538             ANTILOG (A+1 S.D. OF A)=    3.20897
SLOPE B =   -0.13013E+00           STD.DEV. B =      0.49067E-02
CORRELATION R =   0.993664
VARIANCE Y =     0.47163E-01
RESIDUAL VARIANCE OF Y = 0.66208E-03
RESIDUAL SUM OF SQUARES= 0.59587E-02
SUM XOBS. =    71.50               SUM XOBS. SQARED =    492.250
SUM YLOG. =   -4.0952              SUM YLOG. SQARED =    1.9963
SUM XOBS.*YLOG. =  -30.19765


WEIGHTED REGRESSION
MAXIMUM LIKELIHOOD METHOD    10 ITERATIONS

    YO =   3.130057                KR =    0.13379

MEAN X=  5.65282 MEAN Y= -0.260724 MEAN (RF)=   0.54862
Y INTERCEPT C =     0.49555E+00    STD.DEV. C =    0.29025E-01
ANTILOG C =    3.130057
YO+1STD.DEV. =     3.34639         STD.DEV. YO =   0.21634
SLOPE D =  -0.13379                STD.DEV. D =    0.49879E-02
CORRELATION R =   0.999313
VARIANCE OF Y =     0.47501E-02
RESIDUAL VARIANCE OF Y = 0.65210E-04
STANDARD DEVIATION OF MEASUREMENT =    0.80753E-02
RESIDUAL SUM OF SQUARES= 0.58689E-03
SUM W*XOBS.=    7.77306            SUM W*XOBS. SQARED=  46.56076
SUM W*YLOG.=   -0.35852E+00        SUM W*YLOG. SQARED=   0.14098E+00
SUM W*XOBS.*YLOG.=  -2.37729       SUM OF WEIGHTS=   1.37508
SUM(WXX)=  2.62105   SUM(WYY)=  0.47501E-01   SUM(WXY)=   -0.35066E+00
RESIDUAL SUM OF SQUARES/SUM OF WEIGHTS =     0.42681E-03
```

Figure 10. Representative output page 1 for PAGE-PACK program RFT1. The important practical output parameters K_R, Y_o, their standard deviations and the correlation coefficient of the Ferguson plot obtained by a weighted linear regression analysis are marked.

Output 2

```
THURSDAY              MARCH  8, 1984

SYSTEM 10013.000     CROSSLINKING    2.00      PROTEIN  100.00
JOINT 95 PERCENT C.L. FOR YO AND KR

F =   4.26

     0.49555E+00   -0.13379    5.65282    -0.26072
   11   1.37508    0.65210E-04    2.62105

SLOPE                YO-LOWER             YO-UPPER
   -0.14834           3.78290              3.78290
   -0.14106           3.30588              3.58171
   -0.13379           2.98853              3.27828
   -0.12651           2.73535              2.96358
   -0.11923           2.58988              2.58988
```

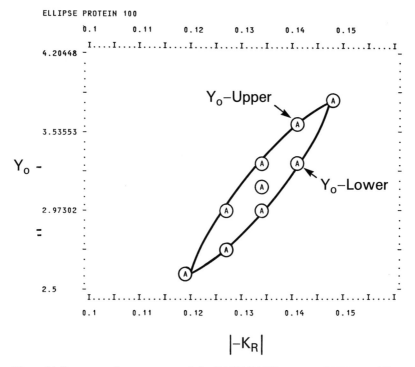

Figure 11. Representative output page 2 for PAGE-PACK program RFT1, providing numerically the joint 95% confidence envelope of K_R (designated as slope) and Y_o. A manual plot of the envelope is also shown.

Output 3

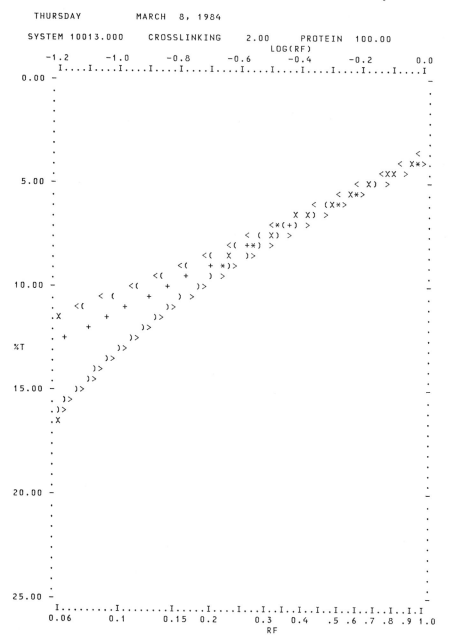

Figure 12. Representative output page 3 for PAGE-PACK program RFT1, providing a Ferguson plot with (inner) 95% confidence envelopes around the line and (outer) 95% confidence envelopes for a single data point along the line.

Input File DATA03

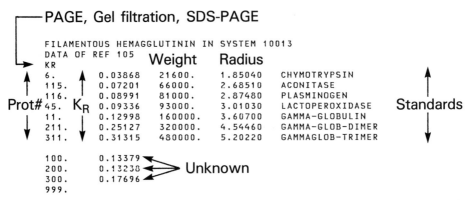

Figure 13. Representative PAGE-PACK input file DATA03 for program GIANTRUN (top) and the listing of the molecular weight standards on page 1 of the output of program GIANTRUN.

Program Giantrun

```
THURSDAY            MARCH  8, 1984

FILAMENTOUS HEMAGGLUTININ IN SYSTEM 10013
DATA OF REF 105

RUN DATA

X = SQRT    KR          Y = RADIUS

  0.19667E+00             0.18504E+01    1
  0.26835E+00             0.26851E+01    2
  0.29985E+00             0.28748E+01    3
  0.30555E+00             0.30103E+01    4
  0.36053E+00             0.36070E+01    5
  0.50127E+00             0.45446E+01    6
  0.55960E+00             0.52022E+01    7


NO SIGNIFICANT NONLINEARITY IN PARABOLIC TEST

UNWEIGHTED LINEAR REGRESSION OF  Y  ON X

         Y =   0.26015E+00 + ( 0.88102E+01) * X

MEAN X = 0.35597E+00          MEAN Y = 0.33963E+01
Y-INTERCEPT A1= 0.26015E+00   STD.DEV. A1 =    0.14430E+00
SLOPE B1 =    0.88102E+01     STD.DEV. B1 =    0.38402E+00
CORRELATION R = 0.99528
RESIDUAL VARIANCE OF Y =    0.14945E-01
SIGMA-SUB-Y = 0.12225E+00      SUM(XX) = 0.10134E+00



UNWEIGHTED LINEAR REGRESSION OF  X  ON  Y

         X = -0.25900E-01 + ( 0.11244E+00) * Y

X-INTERCEPT A2 =-0.25900E-01   STD.DEV. A2 = 0.17445E-01
SLOPE B2 = 0.11244E+00         STD.DEV. B2 = 0.49009E-02
RESIDUAL VARIANCE OF X =     0.19073E-03
SIGMA-SUB-X = 0.13811E-01      SUM(YY) = 0.79409E+01


STUDENT'S T VALUE FOR  5 DEGREES OF FREEDOM =   2.57


INTERPOLATION OF UNKNOWNS USING VALUES FROM THE REGRESSION OF Y ON X
```

PROTEIN	KR	RADIUS	UPPER LIMIT	LOWER LIMIT	MOLECULAR WEIGHT	UPPER LIMIT	LOWER LIMIT
100.0	0.13379	3.4827	3.8181	3.1473	144002.	189741.	106276.
200.0	0.13238	3.4657	3.8010	3.1303	141900.	187207.	104565.
300.0	0.17696	3.9663	4.3076	3.6250	212708.	272471.	162392.

Figure 14. Representative PAGE-PACK output page 2 of program GIANTRUN, providing molecular radii and weights for the unknowns, together with their 95% confidence limits, assuming sphericity of unknowns and standards. The analogous data for random-coiled unknowns and standards (not shown) are provided by page 4 of the GIANTRUN output.

```
THURSDAY              MARCH  8, 1984           Output 3
FILAMENTOUS HEMAGGLUTININ IN SYSTEM 10013
DATA OF REF 105

                              RADIUS
          0.0         1.0         2.0         3.0         4.0         5.0         6.0
          I....I....I....I....I....I....I....I....I....I....I....I....I
  0.000 -(    +   )>                                                    -
         .X    +    )>
         .<(    +    )>
         .  <(    +    )>
         .   <(   +   ) >
  0.050 -   < (    +   ) >                                              -
         .    < (    +   )>
         .    < (    +   )>
         .     < (    +   )>
         .      < (   +  ) >
  0.100 -        < ( +   ) >                                            -
         .        < ( +   )>
         .         < ( +   )>
         .          < ( +   )>
         .          < ( + ) >
  0.150 -           <(    +  ) >                                        -
         .            <(    +  ) >
         .            <(    + )>
         .            <( +    )>
         .            <( + )  >
  0.200 -            <(*+  )  >                                         -
         .            < ( + ) >
 SQRT    .            < ( + ) >
 KR      .            < ( + )>
         .            < ( +) >
  0.250 -            < ( +) >                                          -
         .            < ( +) >
         .            < (+*) >
         .            < (+ ) >
         .            < (+ ) >
  0.300 -            < (X) >                                           -
         .            < (X) >
         .            < (+) >
         .            < ( +) >
         .            < ( +) >
  0.350 -            < (+ ) >                                          -
         .            < (+ X >
         .            < (+)  >
         .            < (+) >
         .            < (+) >
  0.400 -            <  (+) >                                          -
         .            < ( +) >
         .            < ( +) >
         .            < (+ ) >
         .            < (+ ) >
  0.450 -            < (+ ) >                                          -
         .            < (+ ) >
         .            < ( + ) >
         .            < ( + )>
         .            < ( + )>
  0.500 -            < X + )>                                          -
          I....I....I....I....I....I....I....I....I....I....I....I....I
          0.0         1.0         2.0         3.0         4.0         5.0         6.0


                               RADIUS
```

Figure 15. Representative PAGE-PACK output page 3 of program GIANTRUN, providing a plot of $(K_R)^{1/2}$ *vs* $\overline{R}$, assuming sphericity of unknowns and standards. Inner and outer 95% confidence envelopes should be interpreted as in Fig.12. Output page 5 (not shown) provides the analogous plot of K_R *vs* MW, assuming a random-coiled conformation of unknowns and standards.

Program Charge

```
10013.    100.       .085        .0100      0.
100.      3.130057   3.4827
10013.    100.       .085        .0100      0.
200.      2.391415   3.4657
10013.    100.       .085        .0100      0.
300.      2.887876   3.9663
```

Input File
Data04

Output

```
SYSTEM 10013.00 CROSSLINKING**** PROTEIN   100.0 YO =3.130057

RM(1,9)    IONIC STRENGTH  TEMP.   MU(SODIUM+)       MU(FRONT)
 0.085        0.0100         0.     0.000248          0.21114E-04


FREE MOBILITY =                                  0.6608781E-04   ←——————
RADIUS =                                         0.3482701E-06
COUNTERION RADIUS =                              0.2500000E-07
DEBYE-HUCKEL RECIPROCAL THICKNESS =              0.3248667E+07
HENRY'S FUNCTION OF   1.13141 = X1 =             0.1032998E+01
CHARGE (COULOMBS/MOLECULE) = Q =                 0.1540183E-10
VALENCE (NET PROTONS/MOLECULE) = V1 =            0.9620171E+01   ←——————


MOBILITY AT I=.1   =                             0.3515821E-04
HENRY'S FUNCTION OF   3.57784 = X1 =             0.1111362E+01
```

Figure 16. Representative PAGE-PACK output page of program CHARGE, which computes on the basis of Y_o, K_R and $\overline{R}$ (input file DATA04) and the other input parameters shown (input file RMDATA, not shown) the free mobility and net charge (valence) of the unknown.

Input File DATA06

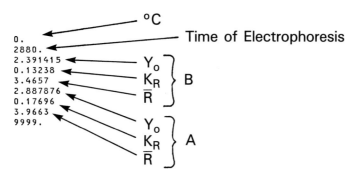

```
0.
2880.
2.391415
0.13238
3.4657
2.887876
0.17696
3.9663
9999.
```

Figure 17. Representative PAGE-PACK input page of datafile DATA06 for program TOPT.

Program TOPT

Output 1

```
TEMPERATURE =   0.0     TIME = 2880.
SPECIES A: Y ZERO = 2.391415    KR =0.13238    RADIUS = 3.46570
SPECIES B: Y ZERO = 2.887876    KR =0.17696    RADIUS = 3.96630
V1 = 0.00000     V2 = 0.00000

GEL CONCENTRATION WHEN MOBILITY OF A = MOBILITY OF B    1.837677
GEL CONCENTRATION FOR MAXIMAL SEPARATION    4.665239
GEL CONCENTRATION FOR OPTIMAL RESOLUTION    7.492802   ◄──────────
GEL CONCENTRATION WHEN SPECIES A UNSTACKS    2.860364
GEL CONCENTRATION WHEN SPECIES B UNSTACKS    2.602727

ANOTHER T-MAX AND T-OPT OCCURS WHEN T = ZERO

SIZE AND CHARGE SEPARATION ANTAGONISTIC; WE SUGGEST A CHANGE OF
PH, ISOELECTRIC FOCUSING, OR ISOTACHOPHORESIS UNLESS YOU ARE
DEALING WITH DIMERS OR OLIGOMERS.
```

GEL CONCENTRATION	SEPARATION	RESOLUTION	MOBILITY OF A	MOBILITY OF B
0.000000	-.496461	-1.087041	2.39141	2.88788
4.665239	0.145323	0.714342	0.57686	0.43154
7.492802	0.107296	0.850016	0.24365	0.13635
2.860364	0.099655	0.359430	1.00000	0.90034
2.602727	0.081697	0.281829	1.08170	1.00000
30.000000	0.000241	0.068598	0.00026	0.00001

Figure 18. Representative PAGE-PACK output page 2 of program TOPT, providing the optimally resolving gel concentration for a set of 2 species, A and B.

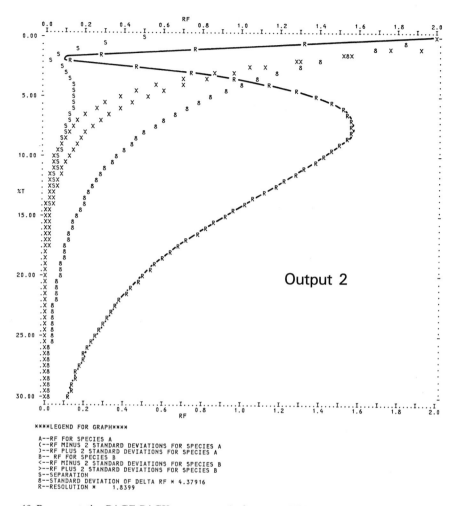

Figure 19. Representative PAGE-PACK output page 3 of program TOPT, providing a plot of resolution (designated by R) and separation (designated by S) between species A and B *vs* %T.

4.13   Preparative Apparatus for Gel Slice Extraction and Concentration

    The simplest and generally most useful method for preparative PAGE and EF consists of gel slicing, pooling of the slices with the protein of interest, and electrophoretic extraction and concentration by MBE [64],[65]. The stacked protein is allowed to migrate into a collection cup with semipermeable membranous floor

(Fig.20). Depending on the diameter of the tube in the separation experiment, and on the choice of PAGE, EF or MBE as the separation method, load capacity is microgram- or milligram preparative. Representative load capacities for the 3 methods are 0.4, 8 and 80 mg/cm^2 maximally. Apart from tubes enlarged at the mouth (to hold the desired number of gel slices) and from the collection cup (Fig.20) ordinary tube gel apparatus is being used (Fig.5). Models I to III with thermostated upper

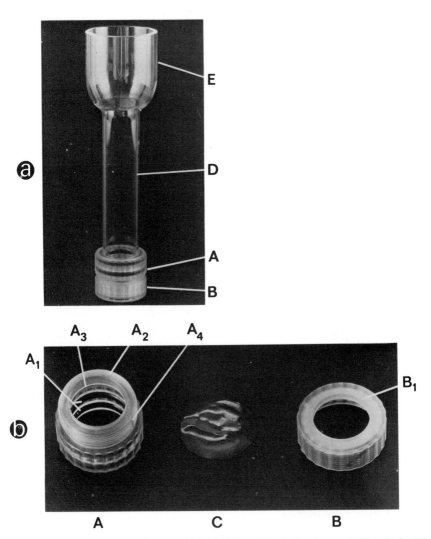

Figure 20. Collection cup attached to gel tubes [65]. Panel (a): Collection cup (AB); gel tube (D); reservoir for loading gel slices (E). Panel (b): Collection cup: A threaded assembly of male Part A and female Part B. Part A features double O-ring (A1), a surface to support the dialysis membrane (A2), inner ledge (A3, faces the O-rings and therefore not visible in this photograph) and air release groove (A4). Part B contains the freely rotating Teflon washer (B1, shown in place, visible through the polycarbonate as a white ring). A dialysis membrane disc (Part C) fits onto surface A2.

buffer chambers, and gels of 18 mm diameter are applicable. The only crucial element in this design, the collection cup (Fig.20), is readily constructed but not as yet commercially available.

4.14 Preparative Elution-PAGE Apparatus

Microgram-preparative elution-PAGE can be carried out by attaching a flow chamber to the bottom of a 6 to 18 mm ID gel in ordinary tube apparatus. Such flow chambers are commercially available (Hoefer No. GT-15E and -6E, or Savant No. PAG-15).

For milligram-preparative PAGE or gram-preparative MBE, elution-PAGE apparatus with 16 or more cm² of gel surface area [18] is required. A suitable apparatus is available (Polyprep-200, Buchler, No. 3-1780), but needs to be modified by addition of accessories such as a manometer tube, Mariotte bottles with 2 L volume and a heavy ringstand mounted on a Lab-jack (Fig.21) ([18],[66],[67], Appendix II or III

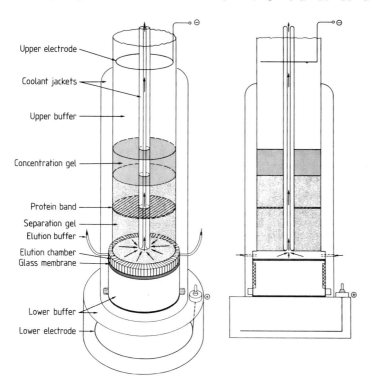

Upper electrode

Coolant jackets

Upper buffer

Concentration gel

Protein band

Separation gel

Elution buffer

Elution chamber

Glass membrane

Lower buffer

Lower electrode

Figure 21. Schematic diagram of a preparative elution-PAGE apparatus with up to 20 cm² gel surface area. The gel column part of the apparatus is the Buchler Polyprep-200 model. For design principles see [167], for performance characteristics [50], for procedure [50] (Appendix D) and [18].

of [50]). Procedurally, this type of preparative apparatus is more difficult and hazardous than the one described in Section 4.13). Although continuous elution flow rate deceleration is possible [68], the dilution of the eluted protein is considerable, necessitating a concentration step such as MBE (Section 4.13). Therefore in general, preparative apparatus of the type described in Section 4.13 is preferable, unless proteins are separated at the gram level by MBE [69].

5   Polyacrylamide

To date, crosslinked polyacrylamide and agarose (Chapter 6) are the sole matrix materials for gel electrophoresis providing a wide continuum of pore sizes for exerting molecular sieving effects on charged molecular species. In the case of crosslinked polyacrylamide, these pore sizes allow for effective molecular sieving of species ranging from a few hundred to several million in molecular weight. A thorough knowledge of this material is a prerequisite for the separation of charged molecular species by Quantitative PAGE. The term "pore size" here designates nothing more than the resistance to migration exerted by a polymer network on a randomly shaped molecule with molecular dimensions substantially less than those of the polymer fiber. The term pore thus is not associated with any geometric picture or model but is being used as a measure of resistance to migration in a random network of fibers, where the length of the polymer per volume defines the restrictiveness of the gel [70]. Calculated values of median pore size (nm) have only didactic and comparative significance (Fig.3 of [72]). Equally, the concepts of a 0-dimensional (0-D), and 1-dimensional (1-D) gel [70] represent limiting cases which are merely approximated by various types of real gels. The 0-D gel element represents a gel whose fiber length approximates zero (a "point-gel"). Supercoiling of linear polymers which gives rise to "ropes" of 10 to 30 nm fiber radius, and which is brought about by a high degree of crosslinking [73] of polyacrylamide, appears to yield gels with "0-D" character.The 1-D gel, by contrast, has "infinite" fiber length and a negligible number of gel fiber endings. Conventionally (2-5% Bis) crosslinked polyacrylamide appears to have predominantly 1-D gel character with a 1 to 2 nm fiber radius. To the degree that real gels approximate either the 0-D or the 1-D type, there exist practically meaningful corollaries with regard to the kind of function which describes the retardation of migration in dependence of the molecular size of the migrating particle (Figs.14 to 16 of [42]).

A crosslinked polyacrylamide gel is formed by the reaction shown in Fig.22. This reaction proceeds via a free-radical mechanism and therefore depends in its rate and extent on 1) the net concentration of available free radicals. It also depends, like any other reaction, on 2) temperature and 3) on reagent choice and purity. All of these parameters need to be controlled to produce a reproducible gel, i.e. reproducible migration rates of charged molecular species on which Quantitative PAGE rests. We will consider now these various parameters individually, then 4) discuss the practical ways of making gels with different pore sizes, and 5) the general procedure for making polyacrylamide gels.

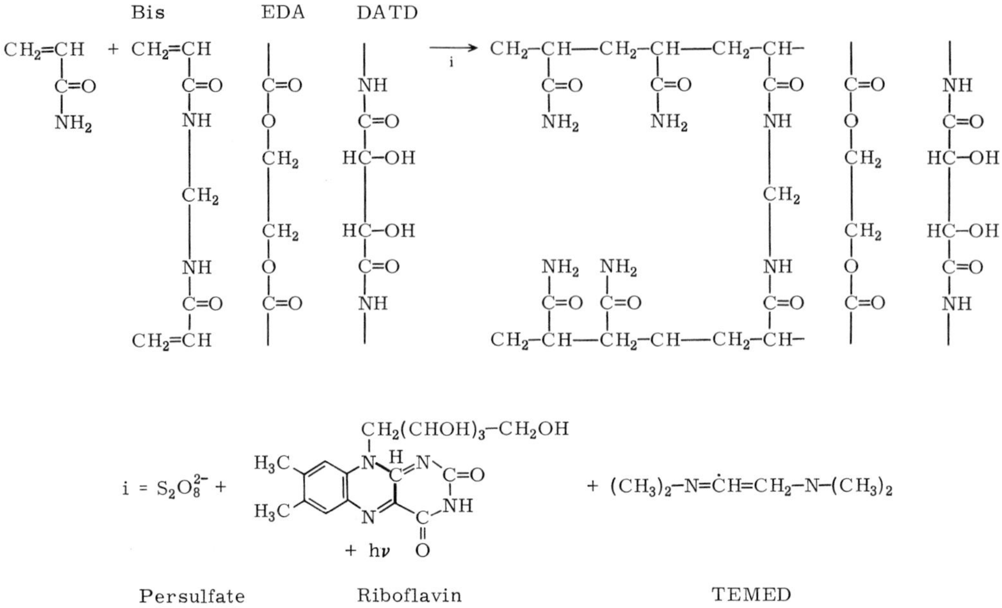

Bis   EDA   DATD

Figure 22. The polymerization reaction which forms a crosslinked polyacrylamide gel. The structure on the left is acrylamide monomer. Three alternative crosslinking agents are shown: N,N'-methylenebisacrylamide (Bis), ethylenediacrylate (EDA) and N,N'-diallyltartardiamide (DATD). Acrylamide and one of these crosslinking agents react to form crosslinked polymer chains of the type shown. The reaction is catalyzed by initiators (i), of which the three jointly used ones in the procedures provided here are shown in the bottom panel.

5.1   Net Free Radical Concentration

The net free radical concentration depends on the chemical nature of the free radical donors (initiators) and their concentrations. It also depends on the concentration of inhibitors (free radical captors). Both of these concentrations depend also on pH and on the rates with which free electron donors decay in aqueous solution, producing peroxides, which again can act as initiators in proportion to their concentration. In the practice of PAGE, the polymerization reaction can be catalyzed to proceed at an acceptably rapid rate by use of only 3 species of initiator: Persulfate, illuminated riboflavin and N,N,N',N'-tetramethylethylenediamine, abbreviated as KP (potassium persulfate), RN and TEMED, respectively. KP is preferred to the ammonium salt when the reaction takes place in buffers above pH 7, to prevent the "hidden" buffering of the system by ammonia (pK 9.1). As a first approximation, the products of the concentrations of KP, RN and TEMED, raised to the appropriate power (a,b and c, respectively) govern the degree of completion ("*% conversion*") and the rate of the polymerization [74]. That means that concentrations of one of the 3 initiator species can be increased at the expense of decreasing another. Thus, one may for instance in systems of positive polarity (protein positively charged) decide to lower persulfate concentrations and increase TEMED, since the former is negatively charged and migrates continuously through the protein zone, while the other is positively charged and migrates out of and away from the protein zone. Such interchangeability of initiators assumes, of course, that exponents a, b and c equal 1. In reality, this is not the case. In an acidic system where data are available [74] it appears that a, b and c are close to 0.2, 0.8 and 0.8, suggesting, in accordance with common experience, that KP is relatively inefficient in acid systems, while riboflavin and TEMED are more efficient in catalyzing the polymerization reaction. Based on experience, we would guess that in basic systems the situation is reversed, i.e. a, b and c would be 0.8, 0.2 and 0.8, respectively, or in other words, that KP is significantly more efficient as an initiator than RN.

The choice of free radical donor concentrations to give a high degree of conversion of monomers to polymer is, however, not sufficient to also provide a mechanically stable, homogeneous gel. A low rate of chain initiation and chain termination to give long, high molecular weight chains (a high *average chain length*) is equally important. While high net initiator concentrations promote a high rate of chain initiation and termination, a high % conversion and low average chain length, a low initiator concentration promotes a low degree of chain initiation and termination, a low % conversion and a high average chain length. Thus, the desired high % conversion and high average chain length originate in antagonistic fashion, the former being favored by high free radical concentrations and the latter by low concentrations. In view of this antagonism between the requirements for a high % conversion and a high average chain length it is important, in practice, to make an attempt to arrive at the minimum initiator concentrations that will give a reasonably high % conversion, i.e. better than 95%. Since the methods available for measuring % conversion are either very laborious [74] or require expensive instrumentation (e.g. [75]), a practical guide to such barely adequate initiator concentrations is the polymerization rate. If a crosslinked gel is formed within 1–2 minutes, initiator concentrations are usually too high, the average chain length too low, and the mechanical properties of the gel therefore poor (cracks, airbubbles, inhomogeneities). If a 6 mm diameter gel forms in more than 30 minutes, it is usually poorly converted, particularly at or near the surface where oxygen inhibition (see below) is high; the mechanical properties of such a gel are also poor, if the nominal gel concentration is low. For gels of 18 mm diameter, the same applies for < 60 min polymerization time. At any gel concentration, such gels are useless for Quantitative PAGE since their effective gel concentrations are unknown and highly variable (variance of % conversion increases with decreasing % conversion [74]). The most useful 6 mm gel is formed under any set of conditions within 10 ± 2 min or twice that for an 18 mm gel. It can be achieved at any gel concentration either by selection of minimum initiator concentrations, or by using high initiator concentrations in conjunction with inhibitor concentrations capable of reducing the polymerization rate to the required value. It is obvious, considering the mass law, that polymerization proceeds more rapidly at higher gel concentrations, requiring lower catalyst concentrations, and *vice versa*.

Net free radical concentrations in the polymerization reaction depend on free radical captors (inhibitors) as much as on free radical donors (initiators). In the very first years of PAGE [24], K-ferricyanide was added to a polymerization mixture catalyzed at high pH by persulfate and TEMED alone, to reduce the polymerization rate. Na-sulfite has been used for the same purpose [76]. The most important inhibitor of the reaction is atmospheric oxygen. For that reason, polymerization mixtures need to be deaerated to a reproducible level, to achieve a reproducible gel. The control of this deaeration, to provide a constant partial pressure of oxygen, may be achieved through temperature control of the polymerization mixture during evacuation. Since the minimum pressure at any temperature is the water pressure at that temperature, one can apply a very efficient pumping system to a perfectly thermostated polymerization mixture without being able to reduce the net pressure below the partial pressure of water. In the same way, one can evacuate a polymerization mixture by water aspirator and obtain a reproducible degree of deaeration as long as the water temperature is constant. An experimentally easier but more expensive approach is to regulate the deaeration pressure over the polymerization mixture electronically [49]. Here, the selected pressure is usually 10 mm Hg at 0 °C, or 20 at 25 °C, i.e. it is selected above the water pressure at these temperatures, making it unnecessary to accurately thermostat the polymerization mixtures. As yet another approach is to saturate the thermostated polymerization mixture with an inert gas such as nitrogen or argon, thus displacing the dissolved air. Argon is preferable since it it heavier than air, thus protecting the polymerization mixture from re-oxygenation, once the passage of the inert gas through the mixture is terminated. Since displacement of air is not instantaneous, passage of inert gas for a constant time at a constant rate is advisable. It is easily achieved by means of a gas flow regulator set a 20 mL/min, when the volume of the polymerization mixture is 10 to 20 mL.

In addition to oxygen, the following polymerization inhibitors need to be considered: Aliphatic alcohols, sulfhydryl reagents, and aromatic compounds such as pyridine, its methyl-derivatives picoline and lutidine, and phenols. Protons – low pH – are also inhibitory. Finally, some apparatus materials can act as polymerization inhibitors – soft (soda) glass, Plexiglas, Tygon but not hydrophobic materials such as polypropylene or Teflon.

Correspondingly, there are also "hidden" free radical donors that should be considered when polymerizing a gel. Rubber surfaces can act in this manner [77]. Small molecular weight anions, cations and divalent metals appear to be catalytic species, since upon their withdrawal by means of 8% crosslinked Dowex 50 and Dowex 1, or by Chelex 100, polymerization fails to proceed.

5.2   Temperature

The polymerization reaction proceeds more easily at room temperature than at or near 0 °C. Nonetheless, at least in application to proteins it is usually carried out at 0 °C. This has at least 2 reasons: First, it is sound to polymerize the gel at the temperature of electrophoresis which in application to proteins is routinely 0 °C for reason of maintaining native properties and conformation as much as possible. This is most evident in preparative elution-PAGE, where one deals with large, heavy blocks of gel which are only tenuously adhering to the glass walls. If one polymerizes at a higher temperature than that used for electrophoresis, these gels separate from the wall; if one polymerizes at the same temperature, they adhere. Thus, thermal contraction of the gel also must contribute to weakening wall adherence of gels even when their mass/surface ratio is less than in the preparative case. Even if this were not the case, analytical conditions should be made the same as preparative ones, in the interest of interchangeability of the characteristic R_f values of proteins.

Secondly, the variance of R_f, and therefore presumably the variance of pore size, is less when one polymerizes at 0 than at 25 °C [74]. Thus, even in SDS-PAGE with fully denatured proteins, there is an advantage to running PAGE in the cold [78]. The reason for this is likely to be related to the increase in average chain length, when the polymerization is allowed to proceed gradually rather than abruptly. The improved average chain length leads to better mechanical gel stability and therefore to a more reproducible gel matrix. This effect is most apparent with DATD crosslinked gels, where under many conditions only 0 °C polymerization gives useful gels.

The necessary quality of temperature control depends on gel concentration: The polymerization reaction is exothermic; thus, the higher the gel concentration, the more heat is evolved and needs to be dissipated. Using the jacketed tube gel apparatus to polymerize the gel, in conjunction with rapid coolant flow and magnetic stirring, the efficiency of heat dissipation is sufficient to polymerize a 40 %T gel of 10 cm length without formation of airbubbles or inhomogeneities in the gel.

5.3   Reagent Choice and Purity

5.3.1   %T, %C

The polymer formed depends on the concentration of monomers and on choice and concentration of crosslinking agent. By a convention introduced by S. Hjerten, the total monomer concentration making up a gel is defined as %T, where % (w/v). Thus,

%T = acrylamide (g) + crosslinking agent (g) / 100 mL.

The % crosslinking is defined as the ratio of the weight of the crosslinking agent to the total monomer weight (not to acrylamide alone). Thus,

%C = crosslinking agent (g) × 100 / %T.

Evidently, the gel is physically characterized by a set of %T, %C values only if the reagent purity of acrylamide and crosslinking agent is perfect and the conversion of monomers to polymer quantitative. In the practice of Quantitative PAGE, where the reproducibility of migration distances at various gel concentrations depends on the reproducibility and accuracy of the gel, it is therefore necessary to apply some care to approximate these two ends.

5.3.2   Acrylamide

Acrylamide in adequate purity can be prepared by two recrystallizations from acetone or chloroform alone (Appendix D of [17]). This step removes the two most important contaminants of acrylamide, viz. polyacrylamide and acrylic acid, to a sufficient degree. Mixed-bed ion exchange chromatography on 8 or 10% cross-linked Dowex 501-D can be used on the acrylamide stock solution after solvent purification to remove acrylic acid quantitatively, but such a step introduces uncertainty into the %T value, since in preparing the resin in the acrylamide stock solution (usually 40 %T) one makes the assumption that the resin volume equals the corresponding volume of water. Also, the anion exchanger part of the resin tends to give off amines, i.e. new uncontrolled impurities. Commercial purified preparations available from many sources (e.g. BioRad, BRL, Polysciences, Serva) are solvent recrystallized, but in addition also purified by many other criteria, e.g. spectral ones, which are irrelevant for most PAGE work and make these preparations more expensive than they would have to be. What these commercial preparations, however, fail to do in most cases is to indicate the date of preparation and the temperature and mode of storage. Both important contaminants, polyacrylamide and acrylic acid, start to form immediately and progressively after purification, particularly when acrylamide is not stored under perfectly dry, cold and dark conditions. Without such storage conditions, and in ignorance about shelf life, there is a good chance that the expense of highly purified commercial acrylamide is largely wasted.

Some explanation should be given, why polyacrylamide and acrylic acid are the most significant contaminants of acrylamide. In the presence of polyacrylamide, the polymerization reaction proceeds more readily, possibly through a nucleation mechanism similar to that in crystallization. Thus, one would lower the catalyst concentrations to adjust to that fact. Since, however, the concentration of polymer varies with shelf-life and between batches, a continuous readjustment of catalyst concentrations would become necessary, which introduces a grave element of pore irreproducibility.

Acrylic acid is important because it may give rise to electroendosmosis and possibly also to binding and retardation of protein cations [79]. Twice acetone recrystallized crude acrylamide in 1 M solution still titrates as an approximately 0.01 N acid without giving rise to measurable electroendosmosis. A 1% value should be considered the maximally tolerable level of acrylic acid in 1M polyacrylamide.

5.3.3 Bis

Bis is the usual crosslinking agent. Its purification requirements are the same as those of acrylamide, except that a single recrystallization appears sufficient. The Bis polymer formed on storage is rather insoluble in water, making it easy to recognize this contaminant. Any acrylamide-Bis stock solution containing insoluble particles should be discarded, or at least not used for Quantitative PAGE (filtration does not help, since %T is modified to an unknown degree).

5.3.4 Alternative Crosslinkers

Although the conventional crosslinking agent is Bis, providing a methylene bridge across 2 acrylamide monomers as a crosslinking structure (Fig.22), many other possible crosslinking agents exist. The methylene bridge in Bis may be substituted by a larger carbon chain; e.g. crosslinkers with 5-12 carbon chains are available (Polysciences No.9816-20, 1495). N,N'-(1,2-dihydroxyethane)bisacrylamide (Polysciences 8031), or N,N'-diallyl-tartardiamide (DATD) (BioRad No.161-0620 or Polysciences No.3058) provide periodate cleavable crosslinks. Moreover, the latter agent allows one to prepare relatively open-pore, elastic and mechanically stable gels [80]. Ethylenediacrylate (Polysciences No.0763) provides alkali dissociable ester crosslinks. N,N'-cystaminebisacrylamide (Polysciences No.9809) provides polymers crosslinked by SS-bonds and therefore dissociable into linear chains by reduction. Attention should be paid to the fact that due to the different molecular weights of these crosslinking agents, their frequency along the linear polyacrylamide chain is not described by the same weight contribution (%C). For instance, 15 %C_{Bis} reflects $15/185 = 0.081$ mol of Bis per ($85/71 = 1.197$ mol of acrylamide) + (0.081 mol Bis) in terms of vinyl groups, the latter being equivalent to 0.162 mol of acrylamide; thus, $0.081/1.440 = 5.6\%$ is the mole fraction of methylene crosslinks per acrylamide monomers. On the other hand, 15 %C_{DATD} reflects $15/228 = 0.066$ mol of DATD per (1.197 mol of acrylamide + 0.132 mol of methacrylate). Assuming acrylamide and methacrylate to be chemically equivalent, this would indicate a mole fraction of $0.066/1.329 = 5.0\%$ crosslinking groups per 100 monomers.

Moreover, the above-made assumption of equivalence is not valid. Methacrylates participate in the polymerization reaction much more sluggishly than acrylamides, leading to the accumulation of pools of methacrylamide and relative enrichment of the growing polymer chain in acrylamide. Toward the end of chain elongation when most of the acrylamide has been used up, the growing chain is correspondingly enriched in methacrylamide. Thus asymmetric polymers are formed [81]. Nonetheless, these polymers may exhibit advantages in mechanical stability as pointed out above.

By converting all %C values to mole fractions in the manner exemplified here, one could devise a unifying way to represent the degree of crosslinking at least for the various acrylamide derivatives. But such a nomenclature is likely to constitute a substantial error source in the preparation of solutions compared to that based on the %T, %C nomenclature. It is doubtlessly for that reason that mole fractions are not being used in this context. For a rational interconvertible use of various crosslinkers, however, mole fractions must be used.

No generalizations concerning the purity of the various crosslinking agents can be made except that all the considerations pointed out for acrylamide apply to them. Some like ethylenediacrylate are very difficult to purify and may be available only in the presence of antioxidants, while commercial DATD in crystalline form appears applicable without further purification.

5.4   Pore Size

5.4.1   Dimensions

Pore size of the gel generated by the polymerization reaction is a function of %T and %C. For any constant value of %C, the retardation coefficient, K_R of a migrating particle (see Section 10.2) is at this time the best measure of effective pore size (see Section 5.4.2). That implies that the determination of pore architecture is limited at present by our knowledge of molecular size, shape, and hydration of "standard" macromolecules. Direct physical characterization of the pore sizes of gels by scanning electronmicroscopy [82],[83] is probably less reliable in this regard because that method alters the water content of the hydrated gel to an unknown degree. Although the scanning electronmicrographs of gel pores stunningly resemble a network, they apparently represent an artifact of preparation since the measured diameters of the visible "pores" of such electronmicrographs are so large that they entirely fail to account for molecular retardation of particles in the size range of proteins.

5.4.2   Pore Size as a Function of %C

Using the K_R values of a few proteins in 10 %T gels of alkaline pH as a measure of pore size at %C_{Bis} values ranging from 2 to 50, it appears that uniformly a biphasic relationship is obtained (Fig.23) ([73], including data of [84] plotted there). Pore size decreases progressively from 2 to 5 %C, while it progressively increases from 5 to 50 %C. In fact, a 50 %C gel does not appear to present an impediment to migration at all for molecules of 1 million molecular weight or less [73]. Presumably, we may interpret this biphasic function in terms of polymer supercoiling: The region of 2 to 5 %C gels represents linear fibers (1-D gels) of increasing density, while the range from 5 to 50 %C represents progressive supercoiling of those fibers, with consequently large interfiber spacings, which behave like "point gels" (0-D gels). In practice, therefore, to achieve gels which are as much as possible "non-restrictive" for the species of interest, one is well advised to turn to the %C range of 10 to 30. Above 30% C_{Bis}, one progressively encounters solubility problems and the need to use mixed solvent systems to maintain the solubility of monomers. Since the mechanical properties of gels – their fragility and strength of wall adherence in particular – deteriorate progressively with increasing %C, the practical choice is always the lowest value of %C that may still be considered to provide "non-restrictiveness".

We should at this point digress and define what we mean by "*non-restrictiveness*" and how it is being measured. First of all, the term is used in quotation marks because it neglects the fact that all gels are necessarily restrictive even to molecules as small as D_2O or urea [85]. Operationally, we use the term to designate a pore size in which the species of interest is retarded to a degree not higher than that of a dye molecule. Specifically, the test system used is a stacking gel operative at high pH

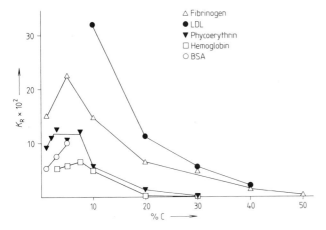

Figure 23. The size of the molecular sieve of polyacrylamide gel as a function of its degree of crosslinking with Bis. The pore size is expressed as the slope of the Ferguson plot for 5 proteins with molecular weights of 1 200 000 (LDL), 340 000 (Fibrinogen), 290 000 (Phycoerythrin), 64 500 (Hemoglobin) and 67 000 (BSA). Note that the gel ceases to be restrictive to migration between 50 and 20 %C_{Bis}, depending on the molecular weight of the particle passing through the gel, and that the most restrictive gel for anyone species is formed at approximately 5 %C_{Bis}. Data of [73].

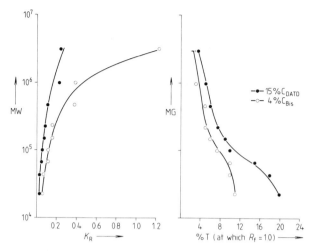

Figure 24. Relative "non-restrictiveness" of two mechanically stable, open-pore gel types as a function of the molecular weight of the protein migrating through the gel. The criterion of "non-restrictiveness" is stacking at pH 10.5 with a trailing ion net mobility of 0.064, 0 °C. Data of [80].

(10.5) and low trailing ion mobility (RM 0.064, where RM designates mobility relative to Na^+) [80]. A gel is considered "non-restrictive" for a molecular species, if that species is capable of migrating with a relative mobility value RM of 0.064 or more at pH 10.5. Fig. 24 depicts some "non-restrictive" gel concentrations determined by that criterion for molecules of varying molecular weight, using gels of 4 %C_{Bis} and 15 %C_{DATD}.

5.4.3   Agarose-polyacrylamide

A different type of "non-restrictive" gel is the so-called "agarose-polyacrylamide copolymer". It provides mechanically stabilized 2 %T gels, or even slightly lower gel concentrations, the restrictiveness of which starts at about one million molecular weight. The combination of the temperature dependent gelation of agarose and the free radical polymerization of crosslinked acrylamide raises experimental problems in the formation of a homogeneous gel. In our experience, neither gelation prior to polymerization [86] nor polymerization prior to gelation [87] but only a synchronized technique (Appendix I of [48]) provides a reasonably reproducible homogeneous gel, albeit at a sacrifice of procedural simplicity. The procedure consists of mixing boiling agarose solution with cold acrylamide stock solution at a 1:1 volume ratio. The resulting mixture at approximately 40 °C is pipetted into the gel tubes, using warm pipets. The tubes are seated in a lower buffer reservoir at room temperature. Coolant flow at 0 °C is initiated together with illumination of the gels. This procedure therefore has the inherent defect that gelation is not allowed to proceed at room temperature overnight as required for homogeneous agarose gels (Chapter

6); the reproducibility of polymerization must be negatively affected by the continuous change of temperature during polymerization.

As noted in the section on gel dryers (Chapter 4), agarose-polyacrylamide "copolymers" are suited for drying into homogeneous gel films, and therefore have a potential importance in the production of prefabricated dried (infinitely stable) sheets of polyacrylamide at all % values. Such dried sheets of 5 % "copolymer" are commercially available (Industrie Biologique Française, 35, Quai du Moulin de Cage, 92 Gennevilliers, France).

5.4.4   Pore Size as a Function of %T

Pore size of long-fiber gels is exponentially related to %T. Thus,
$$\log (M/M_o) = K_R \times \%T,$$
where M designates restricted mobility in a gel, M_o free (unrestricted) mobility, and K_R is the proportionality coefficient. This relation can be derived from the Ogston theory defining the migration of relatively small particles through randomly oriented fiber gels (Fig.15 of [42]) and has been designated as the Ferguson plot in recognition of C. Ferguson who first observed the relationship in starch electrophoresis [88]. Derivation of the relationship from the Ogston theory assumes that molecular retardation ($\log M/M_o$) is proportional to fiber length and to the surface area of the molecular species passing through the gel. The Ferguson plot is usually linearly plotted as $\log (M/M_o$ or $\log (R_f)$ vs %T, keeping %C constant for all %T values. Its slope K_R defines the molecular surface area of the migrating species in the case of a 1-D gel (%C 2 to 5), and molecular volume in the case of a 0-D gel (presumably %C 10 to 50) (see Chapter 10). Conversely, if one had particles of known surface area or known volume, respectively, these could be used to characterize and measure the pore size of 1-D and 0-D gels. In practice, however, such standards are not available, except in a highly approximative way. E.g. fibrinogen, with a 30:1 axial ratio, has been used in an attempt to distinguish between 0-D and 1-D gels [16],[73]. A more practical way exists, however, to measure the thickness of the polyacrylamide fiber (in terms of its radius, r). The plots relating $\sqrt{K_R}$ with molecular radius (for spherical molecules) or with molecular weight (for random coiled molecules) (Chapter 10) yield a value of r(nm) as the negative intercept of these linear plots on the axis of molecular radius or weight. For a 5 %C gel, one predicts a value of r 1.7 nm; very similar values have been found [16],[70]. Similar r values have been found for agarose gel fibers at agarose concentrations of 1 to 10% [21] (see Chapter 6). However, measurements of r on agarose gels at concentrations of 0.035 to 1% [89] by the same method yields much larger fiber radii (see Chapter 7). The interpretation of that discrepancy which is relevant to the pore structure of polyacrylamide is presumably that supercoiled gels (such as agarose or high %C polyacrylamide) provide large pore sizes and exhibit large (10–40 nm) fiber radii at low gel concentrations (i.e in application to large particles) while at high gel concentrations (in application to proteins) the inter- and intra-fiber distances between polymer chains approximate another so that the effective radius becomes that of these polymer chains (i.e. 1–2 nm).

5.5   Polymerization Procedure

The general elements of the polymerization procedure have been designed to provide a maximum of operational simplicity, absence of any elements of art or need for unusual skill, in recognition of the characteristic properties of the polymerization reaction which we have discussed above.

5.5.1   Temperature

The temperature of polymerization is fixed at 0 or 25 °C in the lower buffer reservoir of the jacketed gel tube apparatus, by suitable setting of the temperature of the refrigerated water bath. Coolant flow is maintained at 1 L/min or more. The lower buffer reservoir, filled 3/4 with lower buffer, is stirred magnetically.

5.5.2   Gel Tubes

Gel tubes of the desired diameter (1 mm wall) and 12 cm length or more are cut from Pyrex tubing. They are lightly fire polished without constricting the internal diameter of the tube, cleaned, coated when necessary, sealed with Parafilm and inserted into the upper buffer reservoir.

Coating of the tubes by 1% Gelamide 250 or 1% agarose is required when the gels are mechanically labile, or when, as in EF and MBE, gels are stressed mechanically during operation. The procedures of coating are given in Section 4.2.3, b).

Tubes are sealed with stretched Parafilm as follows. Squares of Parafilm are cut and stretched before winding them tightly and repeatedly (with twisting after application of each layer) over the bottom of the tube.

The upper buffer reservoir is then placed into the lower one, and lower buffer is added to cover as much as possible of the tube length without causing the upper reservoir to float.

5.5.3   Stock Solutions

The following 3 stock solutions are prepared, tested for pH and specific conductance (in case of the gel buffer) and brought to the temperature of polymerization:

1. A 40 %T, 2 %C$_{Bis}$ (or any other desired %C monomer solution.) It is prepared in a beaker, seated in a 25 to 35 °C water bath, with magnetic stirring. After dissolving and attaining the bath temperature, it is transferred to a volumetric flask and diluted to volume. If (polymeric) insoluble particles remain, the solution is discarded or used for non-quantitative purposes only.

2. Gel buffer. Stacking gel buffer is prepared according to the Buffer Tables to provide a low trailing ion mobility [column 2 of the Buffer System Tables (Fig.4; Appendix 1)]. Resolving gel buffer is prepared with a high trailing ion mobility (ibid.). Usually, one constituent is weighed, while the other is being added until the desired pH of the 1:4 diluted solution at 25 °C (shown in column 7 of the Tables, ibid.) is reached. When highly purified reagents are used, one may also prepare the gel buffer gravimetrically, and verify its agreement with the pH given in column 7 of the Tables.

When the Extensive Buffer Systems Output is used (Appendix 2) the stacking gel buffer (4 × Phase BETA) is prepared according to column 2 of Fig.4 of Appendix 2. Here pH (25 °C) and specific conductance, κ, are not corrected for the 4-fold concentration; they lend themselves therefore to a test of the solution at final dilution, but not to the preparation of the buffer, unless one wishes to neglect the pH change concomitant with 4-fold concentration. The trailing ion mobility of the stacking gel buffer is given by column 3 of Fig.3 [designated as *RM* (1,ZETA)]. If other trailing ion mobilities are desired, one choses the desired value from column 1 of Fig.5, Appendix 2. The corresponding buffer composition and pH are given in columns 8-10 of Fig.5, Appendix 2. The buffer composition is given as the final molar concentration of the leading ion (Constituent 2) and the common ion (Constituent 6). In 0 °C systems, the values in column 10 are 0 °C pHs, making it necessary to take a pH measurement at this temperature (check the compatibility of the electrode used with 0 °C, and the 0 °C pH of the standard pH solutions). The resolving gel buffer (4 × Phase GAMMA) is prepared and checked for pH and conductance in analogous fashion.

3. Solution of 2 (out of 3) polymerization initiators. Usually this is prepared from separate initiator stock solutions on the day of polymerization. For many systems, this is a 0.06% KP, 0.002% RN solution, made from 0.6% KP and 0.01% RN stock solutions, by pipetting 10 mL 0.6% KP and 20 mL 0.01% RN into a dark (aluminum foil wrapped) 100 mL volumetric flask and diluting to volume. The shelf life of the 0.6% KP stock solution at 4 °C must not exceed 1 week. The 0.01% stock solution of RN is prepared at 50 °C, cooled to room temperature, transferred to an Al-foil wrapped volumetric flask and brought to volume. It is stable in an amber bottle at 4 °C.

5.5.4 Mixing of the Component Solutions

The 3 component solutions of the polymerization mixture are mixed in a 2:1:1 volume ratio (Fig.25) within an amber bottle containing a magnetic stirring bar. Usually, 2 volumes of the monomer stock solution diluted to a degree so as to provide twice the final monomer concentration (%T) in the gel are used. For example, to make a 9 %T gel, (a) 2.25 mL 40 %T, 2.75 mL water, (b) 2.5 mL gel buffer and (c) 2.5 mL initiator mixture are pipetted into the bottle.

	Lower Gel (3)[b]			Upper Gel (2)			Upper Buffer (1)			Lower Buffer (11)		
	Components/ 100 mL Stock Solution	pH 25°C 1/4 dil.	κ[b] 25°C 1/4 dil.	Components/ 100 mL Stock Solution	pH 25°C 1/4 dil.	κ 25°C 1/4 dil.	Components/ 100 mL Stock Solution	pH 25°C	κ 25°C	Components/ 100 mL/L Stock Solution	pH 25°C	κ 25°C
°C Electrophoresis: 0	%T(A+Bis) a) 8.0 b) 10.0 c) 12.0 d) 14.0 e) 16.0 f) 20.0 g) 24.0 h) 30.0 i) 40 %C(Bis/T)=5			%T=6.25 %C=20								
mmHg (Polymerization): 10												
Volume Ratio of Stock Solutions in Gel: 1												
1	1 N HCl 28.92 mL Tris 11.47 g	8.48	6240	1 N[a] H$_3$PO$_4$ 25.6 mL Tris 1.92 g	6.90	2130	Gly 3.00 g Tris 4.56 g	8.89	287	1 N HCl 50 mL Tris 7.57 g	7.47	4417
1	KP 60 mg RN 2 mg			KP 60 mg RN 2 mg								
	TEMED/100 mL gel a) 250 µL b) 225 µL c) 200 µL d) 175 µL e) 150 µL f) 100 µL g) 75 µL h) 50 µL i) 25 µL			TEMED/100 mL gel 50 µL								

Polyacrylamide Gel System No. 2860

a 1 M = 2 N
b Nomenclature of Appendix 2
c bromphenolblue tracking dye

pH (9)   = 10.00
pH (4)   = 9.63
RM (1.4)[b] = 0.096
RM (2.4)   = 0.790

5.5.5   Cooling

An ice-water bath (or a water bath for 25 °C operation) is prepared in a beaker or within the evacuation (freeze drying) flask of the electronic deaerator (Section 4.7, Fig.7). The amber bottle containing the polymerization mixture is placed into the ice-water bath. The evacuation flask is closed, and the polymerization mixture is evacuated, depending on the temperature of polymerization, at 10 mm Hg (0 °C) or 20 mm Hg (25 °C) for 5 min with magnetic stirring. If an uncontrolled pump is used, its performance should be monitored and kept constant.

5.5.6   Further Procedure

• The requisite volume (e.g. 5 μL) of TEMED is drawn into a Drummond capillary (Thomas No.7695-D16 to -D40) or Hamilton syringe. The capillary is placed horizontally onto the bench. After deaeration, the evacuation flask is opened to atmospheric pressure. Without delay, the capillary is directed vertically into the polymerization mixture and allowed to deliver; the delivery is monitored visually. The mixture is gently swirled a few times, and aliquots are pipetted into the gel tubes. The volume of polymerization mixture per 6 mm ID tube of 12 cm length is 1.2 mL for the resolving gel and 0.5 mL for the stacking gel. If the sample volume exceeds 250 μL, the stacking gel volume is made to at least twice the sample volume. If stacking gels only are polymerized, 1.7 mL are used. The lower buffer reservoir is filled (or emptied) until the top of the gels and the lower buffer are at the same level. The multichamber apparatus (Section 4.5, Fig.6B) requires larger stacking gel volumes since here tubes are longer; they are filled to a height of 1 to 2 cm below the top of the tube.
• Immediately after delivery of the polymerization mixture into the tubes, they are overlayered with water to a height of 2–3 mm, using the overlayering device described in Section 4.9.
• Photoinitiation light sources (Section 4.8, Fig.3 of Appendix 4) are concentrically assembled around the gel tube apparatus at a minimal distance and are turned on for 30 minutes.
• Just prior to electrophoresis the overlayering water is removed from the gel. When mechanically stable resolving gels have been polymerized, this is done by shaking the inverted upper buffer reservoir and wiping of the tops of the tubes with

◁ Figure 25. Representative recipe sheet for polyacrylamide gel electrophoresis in multiphasic buffer systems. The buffer system number and the buffer compositions and buffer properties are derived from the Extensive Buffer Systems Output [14]. Note that two initiator concentrations are kept constant, while the third (TEMED) is varied as a function of %T to achieve a polymerization time of 10 min. κ designates specific conductance (μ ohm^{-1}cm^{-1}).

tissue. When relatively labile resolving gels or stacking gels have been polymerized, the upper buffer reservoir is gently inverted and the overlayering water is adsorbed onto tissue or cotton swabs.

• The upper buffer reservoir is replaced into the lower reservoir and the reservoirs are filled with catholyte and anolyte, respectively, to the identical level.

6   Agarose

It has been known for a long time that agar, and its partially desulfonated derivative agarose, can be used as a matrix for gel electrophoretic separations [90]. But until recently, its use had been nearly entirely supplanted by that of crosslinked polyacrylamide, in view of the suitability of the polyacrylamide pore sizes for the molecular sieving of species with less than 1 million in molecular weight, and in view of its better mechanical stability and adherence to glass walls, as well as the absence of measurable electroendosmosis. The renewal of interest in agarose is due to the fact that means have been found recently to overcome each one of these defects and limitations.

6.1   Pros and Cons

Compared with polyacrylamide, agarose has several advantages and disadvantages as a matrix in gel electrophoresis. The advantages are that 1) it forms a thermally reversible gel. Gelation by mere cooling is evidently a much simpler and less hazardous procedure than a free radical catalyzed polymerization, and should therefore be more highly reproducible; 2) the gel is made of supercoiled polymer strands [91] and thus exhibits large pore sizes, unless very high gel concentrations are employed [21]. (The retarding effect of high concentrations is due to double-helical agarose fibers constituting the supercoil.) Thus, agarose is capable of being "non-restrictive" and applicable in those gel electrophoretic procedures which require the absence of molecular sieving effects – MBE and EF. It also lends itself to the fractionation of large particles such as viruses of 20–50 nm radius [40]; 3) it is preparatively attractive since diffusion out of a large-pore gel is relatively fast. Potentially, separation of molten agarose from proteins may become a convenient preparative procedure.

The disadvantages of agarose compared to polyacrylamide are 1) its decreased mechanical stability and adherence to apparatus walls; 2) its sensitivity to Joule heating; 3) electroendosmosis; 4) binding of positively charged proteins [97]; 5) heterogeneity of composition and batch variability expected from a natural product (e.g. variable content of "agaropectin" of variable composition).

6.2   Types of Agarose

Agarose does not exist – only many agaroses do. These fall into 3 categories: 1) Agaroses graded according to the degree of substitution with negatively charged sulfate and carboxylate groups and therefore the degree of electroendosmosis they exhibit. In descending order of electroendosmosis, these are the Marine Colloids products HEEO, HE, ME and (LE, HGT and HGT[P]), and the Litex products HSA, HSB and HSC. 2) "Zero electroendosmosis agaroses" produced by a) 20-40% addition of a viscous polymer, Clarified Locust Bean Gum (CLBG) (Marine Colloids [92]) or b) chemical substitution of positively charged groups equivalent to the net negative net charge of agaroses ("Agarose IEF", Pharmacia [93], and most likely Litex' HSIF agarose). 3) Hydroxyethylated agaroses with progressively lowered melting points [94]. These agaroses are graded according to % substitution with hydroxyethyl groups and corresponding melting point decrease. The most highly (about 15%) derivatized species is designated as SeaPrep by Marine Colloids; in its properties it is similar to Agarsieve (Litex). SeaPrep agarose gel melts at room temperature. An about 5% hydroxyethylated species (Marine Colloids, SeaPlaque) melts at higher temperatures. This material presumably corresponds to Litex' EasyPlaque and LS-Series.

6.3   Wall Adherence

Agaroses adhere poorly to glass walls compared to polyacrylamide gels. In the past, they have therefore been applied exclusively to horizontal gel slab apparatus where the gel is fully supported, with necessary sacrifice of the many features of gel tube apparatus which are advantageous for Quantitative PAGE and kinetic analyses in EF (see Chapter 4). Only relatively recently has agarose become applicable to apparatus of any geometry. This is due to the finding (C. Saravis, Boston City Hosp., Boston MA, personal communication) that glass surfaces, or even hydrophobic surfaces like Mylar sheets, when coated with agarose adhere firmly to agarose gels. Coating is carried out by filling the tube or immersing the sheet in agarose of type 1 or 2 (Section 6.2) at 60 to 80 °C, and allowing the vertically held tube or sheet to drain onto tissue paper inside an oven maintained at 60 °C.

6.4 Gelation

The gelation of agaroses is best carried out at room temperature overnight. A more rapid gelation at lower temperatures does not yield a homogeneous and reproducible gel. Gel surfaces do not need to be overlayered with water as on polyacrylamide gels. After overnight gelation, gel tubes can be cooled to 0–4 °C without appreciable gel shrinking or other changes in gel structure which would affect molecular sieving.

6.5 Pore Size

Agarose, like highly crosslinked polyacrylamide gel, exhibits "non-restrictive" pore sizes with regard to proteins since its double-helical polymer chains super-coil, i.e. interact to form rope-like, bundled thick chains with radii of the order of 10–20 nm from double-helical fibers of 1–2 nm radius. Supercoiling at low gel concentrations (to about 2%) gives rise to enlarged channels between the "ropes" and consequently large pore sizes. Thus, supercoiled agarose lends itself to the sieving of particles like viruses [40] and large size DNA [98] or to providing a non-restrictive anticonvectant medium for MBE or EF of proteins. In both applications, electroendosmosis-free (type 2) agarose is preferable. Alternatively, electroendosmosis must be measured and deducted from the apparent free mobility values of the particles or proteins under investigation [95],[96].

By contrast, agarose at high gel concentrations (3–10%) exhibits molecular sieving properties relative to proteins similar to crosslinked polyacrylamide gels (2 to 5 %C_{Bis}) [21]. It is assumed that this molecular sieving effect is due to the density and proximity of supercoiled chains, rather than to "un-supercoiling" as had been postulated originally. This re-interpretation is based on the finding [89] that at low gel concentrations applied to fractionation of large (20–50 nm) particles, type 3 agarose is the least restrictive, types 1 and 2 the more restrictive medium. Thus, at high gel concentrations the inter- and intra-supercoil distances appear to become equal, giving rise to a double-helical gel structure with a single interchain distance.

6.6   Electroendosmosis

Measurement of electroendosmosis conventionally employs either the ampho-teric red dye, cyanocobalamin (vitamin B_{12}) [95] or dextran 500 which can be detected by precipitation with ethanol [90]. The use of either one of these reagents is ambiguous since the coordinate covalent linkage of Co in cyanocobalamin is not entirely independent of pH (neutral), and since dextran with a MW of 500 000 is subject to molecular sieving as the agarose concentration is increased. Appropriate methods of correction of mobility values for electroendosmosis are given in [95] and [96]. The commercially used term "$-m_r$" refers to albumin displacement at pH 8.6. It is therefore neither a general measure of electroendosmosis nor a parameter useful for correction. The endosmosis measurement and correction remains an important element of agarose gel electrophoresis, in spite of the existence of agaroses free, or nearly free, of electroendosmosis, since the agarose modifications leading to "0 electroendosmosis", i.e. the addition of CLBG or the introduction of zwitterionic groups lead to new problems which one may wish to avoid.

6.7   Problems Associated with Electroendosmosis-free Agaroses

The addition of CLBG to suppress the electroendosmosis of agarose by an increase in the viscosity is hazardous for 3 reasons: 1) Agarose-CLBG binds to positively charged proteins under the conditions of gel electrophoresis at 0.01 ionic strength [97]. 2) No such binding exists after pre-electrophoresis of agarose-CLBG at 0.1 ionic strength. This treatment presumably removes CLBG from the gel. This should have the effect of re-introducing electroendosmosis, and 3) suggests that preparative agarose gels would lead to contamination with CLBG, even at 0.01 ionic strength. The presence of a non-covalently bound gel component like CLBG which separates during electrophoresis is particularly uncomfortable since its composition and structure are ill-defined and presumably batch-variable.

The presence of covalently bound positive charges neutralizing the negative charges (carboxyl and sulfate) of agarose in the gel is equally uncomfortable since it it known that closely spaced, zwitterionic groups on the gel cause the retardation (binding) of proteins, and in particular positively charged proteins [79]. Whether this problem actually exists in charge-neutralized agaroses or whether charge distances are sufficient to prevent the kind of interactions seen with zwitterionic groups has not as yet been investigated.

6.8 Quantitative Agarose Gel Electrophoresis

The reproducibility of agarose pore sizes should be better than that of polyacrylamide to the degree that it depends on a preformed polymer. Provided that commercial quality control is adequate and gelation conditions constant, agarose gel pore sizes should be reproducible. Since, however, to date quantitative agarose gel electrophoresis has been rarely applied [40],[98], it is not as yet known to what degree batch variability of the original natural product, agar, and of contaminant agaropectin is eliminated by commercial quality control. A further element of pore size irreproducibility affects high gel concentrations only: The high degree of viscosity of agarose solutions in excess of 1 or 2% which makes accurate pipetting or other forms of dispensing very difficult. Pore size reproducibility also depends on the availability of temperature controlled apparatus, in view of the need to prevent any melting or temperature dependent alteration in the gel structure due to Joule heating during electrophoresis. For agarose gel electrophoresis at high (2–10%) gel concentrations, the thermostated PAGE apparatus type 1 has been used [21] in conjunction with very low wattages (current control at 1.2 mA/cm^2 of gel). At lower gel concentrations, a partitioned thermostated horizontal apparatus has been used [22].

Assuming a reproducible pore size, and providing that thermostated apparatus is used for electrophoresis under conditions which produce Joule heat within the limits of the heat dissipation capacity of the apparatus, the mobility measurement in agarose gels poses no problems other than those of rigorous control of the conditions of electrophoresis, and correction for electroendosmosis, if either absolute mobilities (cm^2/s/V) are measured [22] or relative mobilities in discontinuous buffers with negative polarity [97]. In discontinuous buffer systems with positive polarity, proteins may be retarded behind trailing ion mobilities which are attained in polyacrylamide. This retardation has been associated with a component of Iso-Gel, CLBG [97]. Whether or not it occurs in other (CLBG-free) agaroses, remains to be determined. It is also possible, that after removal of an electrophoretically mobile component from CLBG, the binding of cations will cease, just as it can be eliminated by pre-electrophoresis of CLBG-agarose (Iso-Gel) at a sufficient ionic strength [97].

7 Optimization of Stacking pH

Overview

We want to resolve the protein of interest from the mixture in which it is contained. All we have done until now is to define our solvent (Chapter 2), and, as a first step in trying to define the separation pH, we have opted for using a discontinuous buffer system to provide a concentrated starting zone for separation (Chapter 3). We will define in this chapter our first experimental step directed toward finding the optimal separation pH. In summary, it consists of an experimental test of the question what the pH closest to the pI of the protein of interest is at which it remains capable of migrating with a low net mobility (usually taken as 0.050). At such a pH, the net charge ratio between the protein of interest and a protein contaminant from which it is to be separated is maximal (Fig.1), and thus the efficiency of "charge separation" is as great as possible within the limit set by the desired degree of mobility. Experimentally, this test is conducted systematically by setting up a series of stacking gels operative at various pH values. In each case, means must be found to recognize precisely the position of the stack (the moving boundary separating the leading and trailing phases). The most common way to visualize a moving boundary is to add a dye with a net mobility between that of the leading and trailing constituents to the system. If no dye within the required mobility range is available, the position of the stack can be defined by pH analysis of the gel, to find the point of inflection between the pHs of the leading and trailing phases. Alternatively, the stack can be located by direct chemical analysis for the leading or trailing constituent and, in some cases, by inspection of a refractile or stained protein zone exhibiting the characteristic zone sharpness (Fig.3).

Using anyone of these procedures, the stack is located and analyzed for the protein or protein activity of interest. The lowest pH of anionic proteins, or the highest pH of cationic proteins, is found at which all or nearly all of the protein or protein activity migrates within the stack. This pH would represent the optimal pH for separation except for one further qualification. This qualification concerns the question whether, in addition to the pH, the mobility range between the leading and trailing constituents characteristic for the particular selected moving boundary can be narrowed down without retardation of the protein of interest behind the trailing ion of altered net mobility. A series of stacking gels differing in those mobility ranges will be set up to test that question (Chapter 8). Necessarily, an adjustment of mobility range must lead to a change of pH, in a direction opposite to that of the initial pH optimization, but this is usually a slight pH change in the "wrong direction" which does not significantly alter the state of ionization of the protein, the proximity of pH to the pI and thus the resolution based on net charge.

The final, optimal separation pH is then defined as a compromise between the lowest stacking pH and the most specific mobility range between leading and trailing constituents for an ionic protein, or correspondingly the highest pH and most specific mobility range for a cationic protein.

Rationale

The validity of these criteria for optimizing the pH and the net mobility range between leading and trailing ion requires a critical discussion. The rationale for optimization of pH is that proteins are most efficiently separated on the basis of net charge when their net charge ratios are maximal. Net charge ratios between proteins are determined by their titration curves (Fig.1). *A priori,* the logical assumption may be made that charge ratios increase as net charges are decreased (see Fig.4 of [42]). Experimentally, Fig.1 shows such titration curves for trypsin and ovalbumin as an example. The net charge ratios are necessarily maximal (infinite) at the pIs of the two proteins, showing that if we were to separate species only on the basis of net charge, electrofocusing, i.e. separation at the pIs, is necessarily optimal. Since, however, one cannot predict at the beginning of a separation, whether only charge differences, only size differences or differences in both properties prevail in the particular case, a pH must be found which is compatible with migration of the proteins in gel electrophoresis. Fig.1 shows that a pH between the pIs of the two proteins is optimal, and that a pH close to their pIs is relatively favorable by exhibiting a large net charge ratio, while it is at the same time compatible with at least a low mobility. When stacking of the protein of interest is made the tool for the optimization of separation pH, such a "low mobility" is defined for each buffer system as one slightly higher than that of the trailing ion. Since proteins are by-and-large more slowly migrating than small molecular weight trailing ions, the net mobility of the trailing ion must be reduced for that purpose, by selecting a trailing constituent with low ionic mobility and of a pH at which its ionization is repressed (since the net mobility, defined as the product between ionic mobility and degree of dissociation, is thereby reduced). A net mobility of the trailing ion is sought which is equal to or less than that of the protein and thus allows the protein barely to migrate within the range between leading and trailing ion net mobilities, i.e. to be stacked. Since the protein mobility is not known, a low numerical value for the trailing ion mobility is selected, and the inclusion of the protein into the stack at this value of the trailing ion mobility is tested. If necessary, it is then further reduced until the protein stacks. The net mobility values of leading and trailing constituents are expressed relative to the mobility of Na^+ at the particular temperature and ionic strength, to simplify the numbers. The procedure given below will suggest some starting values for the relative net mobility of the trailing ion to be used in conjunction with the various pHs for which stacking of the protein of interest is being tested.

Procedure

The experimental procedure for pH optimization by steady-state stacking rests on a number of conditions and steps that require specification:

- Apparatus
- Gel
- Buffer systems
- Polymerization procedure
- Sample preparation
- Electrophoresis
- Gel removal from the tube
- Detection and quantitation of protein on the gel
- Selection of the optimal buffer system

7.1   Apparatus

A systematic search for an optimal stacking pH usually involves 3 to 10 different pHs and buffer systems. It is obviously impractical to test these buffer systems in 3 to 10 different units of a conventional gel tube apparatus (Fig.5), using a single tube per apparatus. It would be even more impractical, to conduct the test on 3 to 10 different slab apparati, using a single channel in each. What is clearly needed for this purpose is a single multi-chamber apparatus enabling one to conduct the gel electrophoresis in 3 to 10 buffer systems all at once. The simplest design for such apparatus consists of an ordinary PAGE tube apparatus with partitions providing separate upper and lower buffer reservoirs for each gel tube. Such apparatus is not commercially available as yet except as a rather expensive all-Pyrex construction (Fig.6B), which moreover still has some non-optimal design features (see Section 4.5). The most practical approach in securing such multi-chamber apparatus at this time is to cement partitions into commercially available tube apparatus (Fig.6A). The specifications for doing that have been reported [41] and have been critically discussed in Section 4.5.

Since stacking gels are usually mechanically labile, suitable apparatus should provide hydrostatic equilibration of the gels, good wall adherence (Pyrex, coating if necessary) and mechanical gel supports (Nylon mesh, high %T polyacrylamide plugs) (Section 4.2.3). Care should be taken not to exert hydrostatic pressure during manipulations (e.g. while lifting the upper buffer reservoir from the lower one), or suction on those gels during removal of paraffin gel supports after polymerization.

7.2   Gel

7.2.1   Agarose

The gel composition and concentration of stacking gels has only two prerequisites: "Non-restrictiveness" of pore size and mechanical stability. From the viewpoint of "non-restrictiveness", agarose gels appear optimal (see Chapter 6). One would select optimally for this purpose a type of agarose which does not exhibit any measurable degree of electroendosmosis and at concentrations of 1 to 3 % sufficient mechanical stability with absence of measurable sieving effects up to the multi-million MW-range. These gels adhere adequately to glass tubes if they are coated with 1 % agarose solutions by drying for 2 h or more at 60 °C. However, agarose gels have a number of disadvantages as compared to polyacrylamide: a) A much greater sensitivity to Joule heat, requiring one to conduct electrophoresis at 1/2 to 1/4 of the current densities recommended for polyacrylamide; b) greater rigidity, often leading to some wall separation on top of the gel after loading of the protein; c) a lengthy gelation time, since agarose needs to be gelled overnight at ambient temperatures prior to immersion in a 0 °C reservoir electrolyte; d) the possibility of progressive electroendosmosis during electrophoresis due to progressive removal of CLBG or a CLBG component in IsoGel; e) greater procedural complexity when agarose stacking gels are combined with polyacrylamide resolving gels, since the latter gel is formed at 0 °C, then removed from the thermostated reservoir for the gelation of agarose at ambient temperature, on the day before the experiment; f) greater diffusion of the stack making it more difficult to recognize the stack by inspection, particularly when the stack is marked by a tracking dye.

7.2.2   Polyacrylamide

Using polyacrylamide, the degree of "non-restrictiveness" needs to be measured for a protein of particular size as discussed in Section 5.4. As pointed out there, DATD gels appear to be advantageous by combining relative mechanical stability at low %T with a pore size at any one %T value which is larger than that provided by Bis-crosslinked gels. A restriction in the application of DATD crosslinked gels is that they seem to polymerize best at or near 0 °C, and that the pore size cannot be further increased beyond that obtained at 15 %C by an increase in %C (as is the case for Bis up to 50 %C. See Fig.23). One can estimate the maximal gel concentration compatible with "non-restrictiveness" for a given molecular size on the basis of Fig.24, if one keeps in mind the approximate nature of the concept of "non-restrictiveness" of pore size and of the arbitrary way of estimating it discussed in Chapter 5, as well as the fact that Fig.24 was constructed by use of mostly globular proteins [80]. In a practical separation problem, the molecular size range of the sample is usually unknown. It is then probably best to apply the sample to a stacking gel with low trailing ion mobility ($RM = 0.050$) of 5 %T, 15 %C_{DATD} at pH 10 to 11, or at an extreme of acid pH, and to determine experimentally whether upon staining of such a gel all stainable protein is found in the stack. If that is the case, the pore size may be further restricted to make stacking more specific either by increasing %T systematically, or by using 2 to 5 %C_{Bis} gels at the largest %T which does not interfere with the stacking of the protein of interest. By whatever means, the maximal gel concentration should be selected which, within a generous safety margin, is still capable of maintaining all stainable protein within the stack under those conditions. If the protein of interest can be detected by an activity, such choice of a maximally concentrated "non-restrictive" gel may of course be made quantitatively for the protein activity of interest.

If the result of the initial stacking experiment at an extreme of pH in a 5 %T, 15 %C_{DATD} gel is the opposite, i.e the protein fails to stack (and usually migrates as a diffuse zone behind the stack), the experiment should be repeated with the more open 3.125 %T, 20 %C_{Bis} gel, supporting it by Nylon mesh or by a plug of solid gel (e.g. 10 %T, 2 to 5 %C_{Bis}), and taking extreme caution in removing the gel from the tube (see Section 7.7) in order to be able to stain it in an unbroken state.

7.2.3   Agarose-polyacrylamide

"Non-restrictive" gels may also be generated from so-called agarose-polyacrylamide copolymers, if one is alerted to the possible binding problems with agarose referred to in Chapter 6. These "copolymers", formed by superimposition of the gelation of agarose with the polymerization of crosslinked acrylamide, exhibit mechanical stability down to 1.5 or 2.0 %T, when 0.5 to 1.0% agarose are present, while polyacrylamide alone becomes unstable below 3.5 %T, 2 to 5 %C_{Bis}. Various ways to polymerize these mixed gels have been discussed in Chapter 5. In general, application of the mixed gels to stacking of the protein at various pH and under a variety of other conditions appears to be excessively laborious and therefore counterindicated, unless one is forced into their use by a specific inapplicability of the other media to the protein of interest.

7.2.4   Granular Gels

Polyacrylamide may be too restrictive to allow a protein the migration velocity needed to migrate with a mobility more than that of the trailing constituent. Or it may produce too concentrated (sharp) protein zones, leading to aggregation and precipitation of proteins with a tendency to aggregate [99],[100],[105]. Agarose may also not be applicable for any of the reasons listed above, or in view of protein binding. Under those circumstances, a granular matrix which excludes the protein, e.g. Sephadex G-75, can be used to test the stacking of the protein. The procedure of MBE in Sephadex tube gels is analogous to that described for EF in those gels (Section 12.5.2) [54].

7.3   Buffer Systems

The discontinuous buffer systems to be used for the optimization of stacking pH are most easily selected from those of Appendix 1. Only when pHs or net mobility values or buffer constituents or a temperature of 25 °C are needed which are not represented in the 19 selected buffer systems of Appendix 1, should the Extensive Buffer Systems Output (Appendix 2) be searched. This requires a mastery of the terminology of that output which is summarized in Tables 2 and 3 of Appendix 2, and the availability of either the microfiche or the magnetic tape form of the Output, as well as its buffer systems catalog (see Table 1 of Appendix 2 or Fig. 7 of [42]).

The criteria for selection of the systems are:

7.3.1   Operative Stacking pHs

These are taken as the trailing phase pHs (0 °C) in column 1 of the Buffer Systems Tables (Appendix 1 and Fig.4) because the pH of the intermediate phase between leading and trailing ions containing the protein with relatively low net mobility approximates that of the trailing phase rather than that of the leading phase. The trailing phase pHs among the systems (with trailing ion net mobilities of about 0.050) of negative polarity (1 to 14) vary from 10.5 to 5.81; those of positive polarity (15 to 19) vary from 3.5 to 6.70. All of these pHs refer to an ionic strength of 0.01 M and a temperature of 0 °C. If an ionic strength other than 0.01 is required, the pHs corresponding to those in column 1 can be computed by program REFFUB of D. Rodbard, NIH, given in Appendix 7.

7.3.2   Ionic Strength

A second criterion for selection of systems is a low value of the trailing ion mobility (column 2, Fig.4). In all Buffer Systems Tables of Appendix 1, the net mobility of the trailing ion in the trailing phase is approximately 0.050 relative to the mobility of

Na^+ at the same temperature and ionic strength, abbreviated as *RM*. Thus, the initial step in the selection of an initial stacking pH is to select the system with the desired pH in column 1, row 1 of the Tables (Fig.4).

7.4   Polymerization Procedure

Either one of two standard "non-restrictive" gels are usually applied to find the optimal pH, 5 %T, 15 %C_{DATD}if possible and 3.125 %T, 20 %C_{Bis} if necessary. Stacking gels are polymerized in 6 mm ID tubes either in the gel tube apparatus (Section 4.8) or in the multichamber apparatus (Section 4.5), using the procedure described in Section 5.5 (Fig.25) by mixing acrylamide stock solution, gel (leading phase) buffer and catalyst mixture in a 2 : 1 : 1 ratio. In this case, it is easier to prepare a 10 %T stock solution of monomers than to dilute a 40 %T stock solution. As a rule, wall adherence of the DATD crosslinked gels at 0 °C is so good that it is not necessary to coat or mechanically support these gels. However, if 3.125 %T, 20 %C_{Bis} gels are used to test stacking, coating and Nylon mesh support are needed. Alternatively to the use of "standard" gel types for molecules of 0.5 million or less in MW, a specific maximal "non-restrictive" gel concentration may be experimentally determined. This requires construction of a Ferguson plot and extrapolation of the plot to R_f 1.0. The %T at R_f 1.0 gives the maximal "non-restrictive" gel concentration within the definition of the term (see Section 5.4.2).

The leading phase buffer is prepared according to columns 5, 6 and 7 of the Buffer Systems Tables (Fig.4, Appendix 1) at the four-fold concentration appropriate for the mixing ratio suggested above. Both components if pure can be weighed, and the pH of the solution determined. Or, in most practical cases where reagents are impure, the purer of the two components should be weighed and brought to the appropriate pH by the other component, as already pointed out in Section 5.5.3.2. When the complete Extensive Buffer Systems Output is used, the composition of the gel (Phase BETA) buffer is given by column 3 of Fig.4, Appendix 2. Note that here the pH is given at the final concentration in the gel, not at the four-fold concentration, so that the buffers have to be prepared gravimetrically even when reagents are impure, unless one uses program REFFUB to correct the pH to that given by the four-fold concentrate of Phase BETA.

Polymerization catalyst selection can be relaxed in application to stacking gels since in this case an incomplete polymerization or a short average chain length are irrelevant or even favor the "non-restrictiveness" of the gel.

7.5   Sample Preparation

The volume of the sample is irrelevant as long as the stacking gel volume is at least twice the sample volume. However, the buffer composition of the sample should be restricted to the trailing ion and its counter-ion. Within that restriction, it can be made at any pH and ionic strength, under due consideration of Joule heating. Thus, the sample is made at the pH optimal for solubility and activity. The previous practice to prepare the sample in upper buffer defined arbitrarily as a trailing phase in the Extensive Buffer Systems Output [14] or in leading phase (stacking gel buffer) appears unnecessary. That practice was suggested [10] to avoid the formation of stationary (concentration) boundaries at the gel surface. But such boundaries are inconsequential in practice. Furthermore, the preparation of the sample in gel buffer (leading phase) originally suggested on theoretical grounds for systems with strongly acidic and basic leading ions [10], has the practical disadvantage that its pH is frequently detrimental to protein solubility and activities.

The sample should also be provided with sufficient density to allow one to underlayer it under the upper buffer. Usually this is being done by the addition of 10 to 30% sucrose or glycerol. Of course, any other miscible dense non-electrolyte compatible with the native properties of the protein is equally applicable.

If the trailing ion mobility of the moving boundary (stack) allows one to stack a dye or colored protein, this should also be admixed to the sample. Usually, 0.1% aqueous dye (if necessary prepared in a small volume of ethanol and then diluted with water to that concentration) is added to the sample in amounts sufficient to give it a deep coloration. This not only provides a moving boundary marker, but also helps in monitoring sample application visually so as to prevent as much as possible the mixing of the sample with the upper buffer during sample application.

Once the "happiness conditions" for the protein of interest are established (divalent metals, EDTA, REDOX conditions, detergents etc., see Chapter 9), the required agents should also be added to the sample.

Attention should also be paid to the possibility that the dye, or one of the other additives, may react with the sample in a manner which is not completely reversible in the electric field. Albumin, for instance, binds many dyes to give an interaction product with a mobility larger than the protein itself.

7.6   Electrophoresis

Catholyte and anolyte are reservoirs of the trailing and common ion, respectively. They can be prepared at any convenient pH and ionic strength, as long as they contain no other buffer ions. To provide maximal buffering capacity and economy, they are usually prepared at the pK of the trailing or common ion, respectively, using

HCl and KOH to bring them to their pKs (columns 11 and 12, Fig.4). Their ionic strength in the Buffer System Tables is arbitrarily fixed at 0.01 M.

The Parafilm seals are removed from the gel tubes with care not to exert a pull on the gels; this is most easily done by cutting the Parafilm along the gel tube by means of a single edged razor blade, and then peeling off the seal gently. This can be done on the bench, with the upper buffer reservoir inverted, if the stacking gels are mechanically stable. If not, the reservoir is placed horizontally on a cork ring prior to slicing the Parafilm.

The upper buffer reservoir is then placed into the lower reservoir and filled with upper buffer while lower buffer is being poured into the lower reservoir until the levels of upper and lower buffers are the same. This assures hydrostatic equilibration of the gels.

The sample is then pipetted to the surface of the gel, using one of the sample applicators described in Section 4.9 and while visually controlling the smooth flow of sample along the inner tube walls from a height of not more than 1–2 mm, so as to prevent as much as possible any perturbation and mixing of the sample with the upper buffer.

If the sample exceeds a few hundred μL in volume, it is necessary to invert the upper reservoir, adsorb the layering water onto tissues, reposition the reservoir and to pipet the sample into the tube prior to filling the upper buffer reservoir. Then upper buffer can be applied in small droplets to the top of the tubes, using a Pasteur pipet, until the tubes are filled. The upper buffer reservoir is then filled through a funnel directed away from the tubes and with care to avoid any perturbation of the buffer in the tubes. Simultaneously, lower buffer is added to the lower reservoir until hydrostatic equilibration is reached.

Without delay, the current is turned on at 4–8 mA/cm^2 of gel at 0 °C (or twice as much at 25 °C). If the tracking dye contains at least a minor dye component with a mobility between the trailing and leading ions of the selected buffer system(s), this will appear as a characteristically razor-sharp visible zone which does not broaden with increasing migration distance (Fig.3). Once such a visible zone has traversed a distance at least twice that corresponding to the length of the sample zone, electrophoresis may be discontinued.

If no visible marker of the stack is available, one must guess at the appropriate electrophoresis time, with no guide other than experience with electrophoretic rates at 1 mA/tube, and a theoretical relationship. Experience shows that 5 cm of gel are traversed at 0 °C, 4 mA/cm^2 of gel, within 1.5 to 6 hours in various systems operative at 0.015 ionic strength in the resolving gels. Thus, for a 1.7 mL stacking gel, it appears safe to terminate electrophoresis after 1–2 hours in order to test by staining for the appearance of a protein zone exhibiting the characteristic zone sharpness of a stack. The theoretical relation is based on the value of the boundary displacement v, on p.2 of the "Extensive Buffer Systems Output" for each buffer system (Fig.3 of Appendix 2). The position, x (cm), of the stack is related to time, t (min), as $t = 13.1 \, d^2 \, x/v \, i$, where d = diameter of the tube (cm) and i current (mA).

If several gel tubes were run in one buffer system, or if several systems were tested simultaneously by means of a multichamber apparatus (Section 4.5), only a single gel tube is withdrawn at anyone time, and the others are allowed to continue the electrophoresis until they can be conveniently handled. This individual withdrawal of tubes prevents diffusion of the zones prior to fixation or slicing. Since in the multichamber apparatus, single gel tubes run independently of another, their individual withdrawal does not offer any problems. However, when several tubes are run in an ordinary tube apparatus, individual tubes can be withdrawn by a special technique. It consists of placing a rubber stopper (Thomas No. 8823D) into the top of the tube which is to be removed, and a downward pull of that tube through the grommet. As the tube slides out of the grommet, the rubber stopper seals it. A gentle downward pull then tightens the seal of grommet by the stopper.

7.7   Gel Removal from the Tube

Stacking gels are removed by reaming, using a hypodermic needle with water flow as described in Section 4.11.1). All that's peculiar to stacking gels in this regard is the fact that they are relatively mechanically labile in view of their requirement for "non-restrictiveness". When the relatively stable 5 %T, 15 %C_{DATD} gels are used, reaming can be done under a slow, continuous water flow, while holding the gel tube vertically within a few cm height of the bench, to allow the emerging gel to "flow" onto the bench. With 3.125 %T, 20 %C_{Bis} gels the water flow needs to be reduced so as to produce distinct droplets. Reaming is conducted first from one, then the other side of the gel tube. Care is taken to rotate the needle along the circumference of the gel while holding it exactly parallel to the long axis of the gel tube, i.e. to avoid sticking the needle into the gel at any time. The gel after some time emanates spontaneously from the tube and "flows" onto the bench. With stacking gels, one should not blow into the tube to accelerate the process. One should not try picking up the gel from the bench either manually or by forceps, but it should be allowed to slide along the wet bench surface and be pushed to the edge of the bench by means of a curved spatula (scoopula). It is allowed to glide into a vial or tube filled with fixative held against the edge of the bench.

7.8   Location of the Stack

7.8.1   Visible Stack

The position of the stack on the stacking gel may be ascertained by a tracking dye, using as criteria for stacking outstanding band sharpness (of an appearance identical to that of the stained protein stack in Fig.3) and lack of band spreading with migration distance, if at the pH tested a dye can be found which migrates within the mobility range of trailing and leading constituents used. Since there is no way of predicting whether this is the case it is advisable to use mixtures of slowly migrating and rapidly migrating dyes, such as Thionin, Pyronin-Y and methylgreen in systems of positive polarity, and neutral red, RBY (Gelman) and bromphenolblue in systems of negative polarity. Usually some dye component can be located in such mixtures which exhibits tha razor-sharp quality of a stacked band and independence of zone width of migration distance.

7.8.2   Invisible Stack

When no tracking dye of such mobility is available that the position of the stack on the stacking gel can be visualized by its color, three methods are available to locate the stack position:

7.8.2.1   Stack location by analysis for constituents

The stack is defined as the moving boundary across a leading and a trailing constituent. Thus, if a chemical analysis for either the leading or the trailing constituent of a particular buffer system is available, or if either one is available in radioactively labelled form, the stack can be located by analysis for that constituent. Thus, labelled amino acids can be used for that purpose when the trailing constituent is an amino acid. The gel is sliced into 1 mm sections, using a slicer suited for the sectioning of 5 %T, 15 %C_{DATD} gels (the Hoefer electrovibrator slicer, using a rheostat to decrease the vibration frequency; see Section 4.11.3). If a labelled trailing ion is used, the level of radioactivity would be constant in the trailing phase, then decrease in sigmoidal fashion until, in the leading phase, it approaches a blank value. The point of inflection on the sigmoidal function represents the position of the stack.

Among leading constituents, chloride can be precipitated on the gel by $AgNO_3$, sulfate by $BaCl_2$, phosphate by La-acetate. Usually the gel is immersed in 0.1 M solution of these reagents. The sharp edge of the resulting precipitate represents the position of the stack. Often, chemical analysis for trailing or leading constituents can be compatible with later protein fixation and staining, or slicing and quantitation of the protein of interest in the stack since it only requires a very short time.

7.8.2.2   Stack location by pH analysis

Since the pH of the phases separated by a moving boundary varies, it is possible to measure the position of the stack as the point of inflection between the pHs of the leading and trailing phases, providing that the difference in pH between the phases is sufficient to be clearly measurable. If a contact electrode is used to take this measurement (see Section 4.11.10), the same gel can thereafter be subjected to slice excision and assay, or to staining for protein.

7.8.2.3   Stack location by protein staining

The stack may be recognized by the characteristic sharp boundaries and almond-shape after staining for protein (Fig.3) and by the independence of the width of the stained zone of migration distance. Staining of the stack of course cannot indicate whether the protein of interest migrates in the stack, unless this is a substantial part of the sample.

7.8.3   Extended Stack

Since the concentrations inside the stack are regulated, an increase in load leads to the widening of the stack. At high loads of the order of 1 or more mg/cm^2 of gel, the leading and trailing edges of the stack separate, giving the characteristic appearance of a sharply defined "extended stack" [69]. Different dyes may mark the leading and trailing edges of such an extended stack [78]. In that case, the location of the stack position by a single dye leads to an erroneous interpretation of the capacity of the stack to contain the protein of interest. To avoid such an error, gels at high load should be stained for protein to ascertain whether or not an extended stack exists. If so, it may be possible for a given gel concentration to find a pair of tracking dyes to mark leading and trailing edges selectively. Alternatively, the trailing edge can frequently be detected as a refractile or slightly yellowish sharp zone, due to the optical properties of the stacked proteins at their high regulated concentrations.

7.8.4   Procedure of Staining for Protein

An acceptable staining procedure should, first, rely on a universally effective protein fixative, and second, should stain the protein selectively without a stained background. At this time, the fixatives which can be considered universally reliable for proteins of diverse charge densities, carbohydrate content and relative hydrophobicity are trichloroacetic acid (TCA) and protein-crosslinking agents (e.g. 2% glutaraldehyde) exclusively. Neither aqueous or methanolic solutions of acetic acid, in spite of their popularity, are reliable protein fixatives on gels. The concentration of TCA is governed by the second requirement, *viz.* the absence of a stained background. Coomassie Brilliant Blue R-250 [101] and G-250 [102] both are relatively insoluble in 12.5% TCA (and even more so, 10% TCA), and therefore provide saturated solutions of the dye in such media while the proteins avidly take up dye under those conditions. Thus, such saturated solutions are used to stain the protein preferentially over the solvent. The procedure consists of placing the gel into a saturated solution of the dye, and to use a volume of this solution large enough to maintain close to 12.5% TCA after addition of the gel (water) volume.

In the G-250 procedure [102], the 1.7 mL gel is placed into 40 mL 12.5% TCA for 30 minutes. 2 mL of a stable aqueous 0.25% solution of G-250 is added and mixed. After 60 min, the gel is transferred into a 10 mL tube filled with 5% acetic acid. The background-free protein pattern is immediately developed to its full intensity. Minimally 1–2 µg of protein/zone/6 mm gel are required. Optimally, 5–10 µg per zone are used. The color is stable for months if the gels are stored in the dark.

In the R-250 procedure [101], an unstable saturated solution of the dye is prepared by mixing 5 mL stable aqueous 1% R-250 solution with 50 mL 25% TCA; to the resulting also stable solution, 45 mL of water are added with immediate mixing. The resulting saturated solution of R-250 is stable for at most 1 day. The gel is placed into 40 mL of 12.5% TCA for 30 min, then into the same volume of the saturated solution of R-250 for 60 min. After that time, the gel is transferred into a 10 mL tube containing 10% TCA. The color develops immediately against a blank background, as above, but it intensifies in this procedure upon storage in the dark, reaching a maximum after 1–2 weeks of storage in a 10% TCA solution to which R-250 has been added to give a very lightly pale-blue color. Thereafter, the gel shrinks. The sensitivity of the stain prior to storage is approximately the same as that of the G-250 procedure.

Thus, the G-250 procedure is generally preferred, unless one deals with a protein like human chorionic gonadotropin (35% carbohydrate) that resolubilizes with time in the G-250 storage solution (the TCA content of which is approximately 2–3% only), or unless the protein load has been so low that one needs the color intensification with storage time characteristic for the R-250 procedure. TCA soluble peptides may be stained by Coomassie Blue R 250 after crosslinking of the peptides with glutaraldehyde [197]. Gels are fixed in 1.4% picric acid, 20% acetic acid [103], 1.25% glutaraldehyde (40 mL of the mixture/gel) for 30 min. Two milliliter of 0.25% Coomassie Blue R-250 are added and allowed to react for 60 min. Gels are destained in 20% acetic acid in the apparatus for diffusion of SDS from gels [55].

7.8.5   Measurement of Stack Position

The position of the stack is defined in relation to gel length. The ratio of migration distance of the stack/gel length is measured on the unfixed-unstained gel or, better yet, the gel tube prior to removal of the gel. This can be done by ruler or, more conveniently, electronic R_f-measuring device (Section 4.11.5) for a visible stack. In the case of an invisible stack the ratio between the slice number of the stack and the total slices of the gel defines its position, assuming regular gel slices. After fixation and staining, the stack position can be calculated by multiplying the length of the stained gel by the ratio between stack position and gel length obtained on the unstained gel.

Once the position of the stack is defined, it becomes possible to ascertain the presence or abscence of the protein of interest in the stack and to quantitate the protein activity in the stack.

7.9   Quantitation of the Stacked Protein

Stacking of the protein of interest in a particular buffer system, i.e. at a particular trailing phase pH, should be near-quantitative to make that buffer system applicable. However, no strict general rule can be made concerning the value of % protein or % recovery of an activity in the stack. Certainly stacking of a minority of the load of the order of a few percentages is not sufficient to justify the use of a buffer system, since one may be focusing on a minor component and ignore the fact that the bulk of the components fails to stack. Attention must be paid to the values of "% stacked" at the various pHs. If all are low, the problem is likely to consist of one's failure to perform the stacking gel analysis in the presence of the needed "happiness conditions" (see Chapter 9). After remedying that defect, however, recovery of the protein of interest should be at least 60–70% of the load, before the use of a particular buffer system is justified.

The method of quantitation of protein activity in the gel is straightforward: The stack, determined by any one of the methods described above, is excised, eluted and assayed. When a specific color reaction is available the gel can be placed directly into the assay mixture to ascertain qualitatively the presence of the protein of interest in the stack.

By contrast, stack quantitation on the basis of protein staining and densitometry is difficult in view of the fact that the protein concentrations within the stack vastly exceed the range in which color intensity due to protein staining is proportional to protein concentration. Thus, one may determine the unstained protein in the stack after eluting the protein into alkaline buffer (e.g. 0.1 M Na_2CO_3) or 1 % SDS and determining protein on the diluted solution by TCA-Lowry assay (or any other protein assay giving a low blank value with a slice of blank gel electrophoresed under the same conditions) within the known concentration range of the assay. Or, one may elute the stain with pyridine and determine it spectrophotometrically [104].

7.9.1   Stack Analysis by Assay

The percentage of the loaded protein activity located in the stack is determined by excision of the stack, using a razor blade. The excised gel slice carrying the stacked protein is then assayed. If the assay system consists of small molecular weight components, these diffuse into the gel slice and there is no need to elute the protein from the gel. If not, a time period has to be allowed for the diffusion of the protein from the slice. Since the gel is "non-restrictive", and protein concentration is very high, this is usually unproblematic. However, diffusion times between a few hours and overnight should be tested once for each protein, since its diffusion rate depends on its degree of steric restriction through the gel (i.e. the degree of its asymmetry and its size).

7.9.2   Stack Analysis by Staining

After staining, the stacking of the entire protein load is apparent if all of the stained protein resides within the characteristically ultrasharp zone of the stack. The position of that zone (migration distance of the zone relative to gel length) must agree with the value determined for the unstained gel. If the stained gel pattern shows protein migrating behind the stack, it remains unknown whether the protein of interest has been stacked or not. Even if all stainable protein resides within the stack, it remains possible that the protein of interest, in non-stainable concentration, migrates behind the stack. It is therefore necessary, whenever the protein has an activity, to test for the stacking not only of stained protein, but also of the activity.

7.10   Selection of the Optimal Buffer System

Stacking gel electrophoresis is carried out in buffer systems with low mobility of the trailing constituent varying stepwise in operative pH from an extreme of pH toward neutrality or isoelectric point of the species of interest. The choice of the requisite buffer systems has been discussed in Section 7.3, procedure in Section 7.5, evaluation of each gel in Sections 7.8 and 7.9. Usually, only 1 gel per pH is required, and only 1 analysis – that of the gel slice carrying the stack. Several pHs can be analyzed in several analytical apparati, or in a single multichamber apparatus (Section 7.1).

Any pH allowing for recovery of the bulk of the loaded protein is potentially applicable. However, only that pH is selected as being optimal which is closest to neutrality or the isoelectric point of the protein of interest where net charge ratios between proteins, and their separation due to net charge, are maximal.

8 Optimization of Trailing and Leading Ion Mobilities in the Stacking Gel

After having gone through the effort of selecting and running a number of buffer systems, using stacking gels, and after finally having located the one with the optimal pH for separation on the basis of net charge, the natural temptation of anyone interested in resolving his/her protein system and in characterizing and isolating the protein of interest is to proceed immediately to a resolving gel in the selected buffer system. However, this natural temptation should be resisted in favor of further work with stacking gels which, of course, do not provide resolution. Specifically, two kinds of further stacking gel experiments are suggested, before resolving gels should be attempted. The first, discussed in this section, aims at making the stack as narrowly circumscribed and specific for the protein of interest as possible. The second, discussed in Chapter 9, aims at finding conditions in gel electrophoresis which will yield quantitative, or maximal, recovery of the protein or its activity in the gel. These conditions have been termed the "happiness conditions" for the protein of interest.

The reason for attempting to find the maximal value of the trailing ion mobility for the protein of interest is to introduce resolution, as much as possible, into the pre-resolution stage of the separation, i.e. the stacking gel, since such pre-resolution simplifies the later separation problem in the resolving gel. Thus, when the trailing ion mobility is increased so as to just include the protein of interest into the stack, many contaminant proteins will be retarded to migrate as broad, diffuse zones behind the stack. These contaminants, arriving at the surface of the resolving gel, are unable to form a narrow starting zone and consequently will not be giving rise to discrete zones in the resolving gel; they will contribute to a diffuse background stain or activity which does not interfere with the band pattern derived from the stacked proteins.

In the selection of trailing ion mobilities (Section 7.3) the assumption has been made up to this point, that proteins are slowly migrating as compared to buffers making up the moving boundary, and that therefore very low mobility values were required. Now, that an optimal pH for electrophoresis has been found, it is advantageous to abandon that assumption and to ask ourselves what the actually required trailing ion mobility is. Thus, the Buffer Systems Tables (Appendix 1) or the complete "Extensive Buffer Systems Output" will be used to define a number of buffer "subsystems" with stepwise increase in the trailing ion mobility. Electrophoresis will then be carried out in a stacking gel prepared in the desired "subsystems" to find by systematic experiment the maximal value for the trailing ion mobility capable of including the protein of interest into the stack. We define herein as "subsystems" those buffer systems consisting of the same buffer consituents as one of the numbered systems, however at different concentrations and slightly different operative pHs. These "subsystems" are listed in rows 3, 4 and 5 of the Buffer Systems Tables (Fig.4 and Appendix 1) and on p.3 of the "Extensive Buffer Systems Output" for each of the numbered buffer systems (Fig.5 of Appendix 2).

8.1   Selection of "Subsystems" with Increasing Values of the Trailing Ion Mobility

a) The trailing ion mobilities in increasing order from $RM = 0.050$ to 0.250 are listed in column 2 of the Buffer Systems Tables (Fig.4 and Appendix 1). The gel buffer to be prepared in each case is defined by columns 5, 6 and 7, and is prepared as described in Sections 5.5.3.2 and 7.4 by weighing the purer of the 2 gel buffer (leading phase) components and bringing it to the requisite pH by the other.

b) Alternatively, column 1 of the table on p.3 of the "Extensive Buffer Systems Output" for each numbered buffer system (Fig.5 of Appendix 2) lists the values of the trailing ion mobility [RM(1,ZETA)]. For each selected value, column 8 provides the final molar concentration of the leading constituent (Constituent 2) and of the common constituent (Constituent 6) in the stacking gel buffer. Column 5 provides the operative pH [pH(ZETA)] for each subsystem at the operative temperature of the system (usually 0 °C for proteins).

To simplify the calculation of the composition of the stacking ("upper") gel buffer for a subsystem, cognizance should be taken of the fact that the concentration of either Constituent 2 or Constituent 6 often remains constant in all of the subsystems. For that constant concentration, the value given in the Recipe section on page 2 of the Output (Fig.4 of Appendix 2) may be applied directly. In all other cases, the subsystem concentration of a constituent divided by the final molar concentration of the same constituent in the numbered system (column 4 of the Phase Table on p.2 of the Output, Fig.3 of Appendix 2) and multiplied by the g or mL value given for that constituent in the Recipe section on p.2 of the Output (column 4 × Phase 2) provides the g or mL of that constituent in the stacking gel buffer.

The subsystem number by which a subsystem is defined is designated according to the vertical position of the subsystem in either the Buffer System Tables (Appendix 1, Fig.4) or the table on p.3 of the Output, starting the count at the top finite value (other than 0.0), and giving arabic numeral designations to subsystems referring to the trailing phase of stacking gels (Phase ZETA) and roman numeral designations to the trailing phases in the resolving gel (Phase PI).

The subsystems selected from the Buffer Systems Tables are all operative at an ionic strength of the trailing phase of 0.01 M. By contrast, the subsystems selected from the Extensive Buffer Systems Output are operative at variable values of ionic strength, and their operative pHs at 25 °C, where they can be easily tested, remain unknown for 0 °C systems. To unify the operative ionic strength of all subsystems to be tested and to learn the operative pHs [pH(ZETA) values] at 25 °C, computation using the Jovin program (NTIS No.196092) is required (Appendix 2F).

8.2 Computation of the Ionic Strength of a Subsystem of the Extensive Buffer Systems Output

Procedurally, this is extremely easily done, once the Jovin program is installed in a program library and is accessible via a remote terminal. Fig.6, Appendix 2 shows the 6 lines (or cards, if no terminal is available), numbers 60 to 65, required for such a "re-calculation" of a subsystem. Appendix 3 provides the exact significance of each parameter and their positions. However, only very few of these parameters need to be considered in the practice of such re-calculation. In line 60, date and system number (system designation in up to 5 numbers) needs to be stated; in line 61 the polarity of the system (-1 or $+1$) and the temperature (0 or 25 °C); line 63 lists constituents 1,2,3,4,5,6 from left to right. These, of course, are the same constituents as used for the numbered system, and simply need to be read off the top of p.2 of the Output, (Fig.3, Appendix 2), except constituents 4 and 5 which are listed as Cl⁻ or K⁺ on p.1 of the Output (a re-statement of input parameters not shown here). Of sole importance in recalculating a subsystem is line 64 which lists, from left to right, the final molar concentrations in the upper buffer (C1, C6), stacking gel buffer (C2, C6) and resolving gel buffer (C3, C6). These concentrations derive from the subsystems table on p.3 of the Output for those phases that one may wish to recalculate (Fig.5, Appendix 2), and from the top section of the Physical Properties Table on p.2 of the Output (Fig.3, Appendix 2) for those buffer phases that one may not wish to alter relative to the numbered system. If only stacking gel buffers are to be computed for a subsystem, it is of course unnecessary to compute a resolving gel buffer; in that case, line 64 should contain C1-C6, C2-C6, C2-C6.

In line 65 (Fig.6, Appendix 2), the second and third parameter from the left should state the operative resolving pH -1, and the operative resolving pH $+1$.

The input shown in Fig.6, Appendix 2 is then run on the computer, and will provide the ionic strength (ION.STR. on p.2 of the Output, Fig.3, Appendix 2) of phase ZETA. The factor needed to bring this ionic strength to 0.01 M is then applied to C2 and C6 of line 64, Fig.6, Appendix 2. The subsystem is then recomputed, and designated by Subsystem Number($\times$ factor). Thus, e.g. system 1954.3($\times$ 3).IV. would designate a ZETA phase composition corresponding to the third subsystem from the top, with concentrations C2-C6 multiplied by a factor of 3; the resolving gel (Phase PI) would be the 4th, starting with the first number in column 13 (Fig.5, Appendix 2) which is not 0.0.

8.3   Experimental Procedure for Optimizing the Trailing Ion Mobility in the Stacking Gel

Values of the trailing ion net mobility higher than 0.05 are selected from column 2 of the Buffer Systems Table (Fig.4, Appendix 1), or by intrapolation of these values. Alternatively, values of RM (1,ZETA) increasing stepwise above that value previously used to stack the protein (Chapter 7) are selected from p.3 of the "Extensive Buffer Systems Output", column 1 of the Subsystems Table (Fig.5, Appendix 2). Stacking gel buffers corresponding to each selected subsystem are prepared at 0.01 M ionic strength as shown in Sections 8.1 and 8.2. Stacking gels in each of the buffers are prepared, run and evaluated as described in Chapter 7. Since all subsystems have the same constituents, an ordinary gel tube apparatus may be used to conduct electrophoresis in all subsystems simultaneously. The subsystem with the highest value of the trailing ion net mobility is selected which is capable of stacking the protein of interest quantitatively or at a maximal value. This selection has the consequence that the operative pH of stacking, set as closely as possible to the pI of the protein (see Chapter 7), is now slightly moved away from the pI toward one of the extremes of pH. Usually, however, the resulting pH shift in the direction of deteriorating charge fractionation is negligibly small. If it is not, a compromise has to be struck between optimizing the pH and the value of trailing ion net mobility in the trailing phase.

9 Selection of "Happiness Conditions" by Means of Stacking Gels

The experimental selection of an optimal stacking pH (Chapter 7), and the subsequent selection of a maximal value for the trailing ion mobility (Chapter 8) may have yielded stacks which contained the protein of interest near-quantitatively. In that case, gel electrophoresis has already been proved to provide a sufficient set of conditions required for the maintenance of native properties which are designated here as its "happiness conditions". The molecular size of the native species, of course, may still have been in excess of the minimal size required for the expression of an activity, i.e. the size which must be considered of prime interest to biochemistry [3]. It is for that reason only, that a continuation of the systematic stacking gel analysis beyond the optimization of pH, trailing and leading ion mobilities to test the stacking of the native (active) species in disaggregating media may be warranted.

Far more frequent, however, is the need for a test of "happiness conditions" in those cases, where a quantitative stacking of the protein or its activity has *not* been achieved at any pH and at any value of the trailing ion mobility.

Let us assume that no protein activity at all was recovered in the stacks of stacking gels at various pHs and mobility ranges. In that case, a buffer system at high or low pH should be tested for maintenance of the activity in the absence of a gel, by either direct incubation in the buffer or dialysis at a temperature and time representative for a gel electrophoresis experiment. The most extreme pH at which the activity survives should then be applied to stacking gels with systematic variation of "happiness conditions" as described below, until a set of conditions is found which allow for near-quantitative recovery of the activity in the stack.

Assuming that only a small percentage of the activity had been recovered in the stacks at various pHs and trailing ion mobility, it would be inappropriate to proceed with electrophoresis in resolving gels, since that approach would have a good likelihood of aiming at the separation and isolation of a minor active component of the protein mixture, rather than at the major, fully active component detected prior to gel electrophoresis. Again, a systematic scrutiny of "happiness conditions" is indicated in this case also.

The following "happiness conditions" should be considered:

- Temperature
- Ionic strength
- REDOX conditions in the gel
- Solvent
- Charged cofactors such as divalent metals

9.1 Temperature

The activity may have been lost due to thermal denaturation inside of the stack. This temperature arises through Joule heat as a function of the overall current and voltage across the gel, as a function of the radial distance from the center of the gel and as a function of the local high voltage gradient across the stack containing protein at concentrations as high as 20–100 mg/mL [7],[69]. In some cases, such as the preservation of binding activity of a receptor protein [99] the dependence of activity requires a temperature of electrophoresis as low as $-2\,°C$. This required experimentally an insulated gel tube apparatus, a constant temperature bath set at $-4\,°C$ and the addition of 20% glycerol as an anti-freeze to all phases. Dependence of stacking on temperature may be more indirect [105]: It may involve protein aggregation due to Joule heat across and high protein concentrations within the stack, requiring a reduction of regulated current to as little as 1.2 mA/cm² of gel to prevent a reduction in protein mobility in the stacking gel due to large size of its aggregate. Yet another sensitivity on temperature is specific for agarose gels: Depending on the melting temperature, their stability is sensitive to Joule heating, requiring in some cases reduction of regulated current to 2.0 mA/cm² of gel [97].

9.2 Ionic Strength

The usual ionic strength of gel electrophoresis, 0.01 M, is less by one order of magnitude than physiological ionic strength. It is therefore insufficient in many cases for maintaining protein activity and for preventing aggregation or other interactions of the protein due to electrostatic forces. This makes it necessary to stack the protein at various levels of ionic strength, such as 0.03 and 0.10 M in addition to the standard 0.01 M condition. In some cases, such as myosin, gel electrophoresis has to be conducted in 0.6 M salt solution. This is perfectly possible as long as the regulated current is maintained at a sufficiently low value to enable one to dissipate the Joule heat efficiently. Thus, runs in 0.6 M solution take days, and risk time-dependent diffusion of the zones unless the gel is sufficiently restrictive, and the molecule is sufficiently asymmetric, to counteract such diffusion.

Procedurally, stacking gel buffers which give rise to the desired operative ionic strength are prepared by multiplying the g or mL of leading and common ions needed to make 100 mL of the stacking gel buffer (4 × concentrated leading phase) by a factor X needed to raise the ionic strength of 0.01 to the desired value. This gel buffer provides a trailing phase of $X \times 0.01$ ionic strength for the buffer systems of Appendix 1.

If the Extensive Buffer Systems Output is used, the desired concentration factor X is determined which will bring ION STR of Phase ZETA (Fig.3 of Appendix 2) to the desired value. Factor X is then applied to the recipe for the Stacking Gel Buffer given in Fig.4 of Appendix 2. Stacking gels of various ionic strengths can be prepared in this manner and are run simultaneously, using the same catholyte and anolyte. The procedures described in Chapter 7 apply, except for the setting of the current.

If the stacking gels with various ionic strengths are run in individual apparatus, each can be run at a regulated current of 1 mA/tube; the Joule heat then will be maximal at the lowest ionic strength, where it is already known to be dissipated effectively. But setting up different apparatus is not necessary in a case where the constituent compositions of all the gels are the same.

When gels of different ionic strengths are run in the same apparatus, a regulated current level of 1 mA/tube will not provide 1 mA to each tube; rather the current will be higher than 1 in the gels of higher ionic strength, and lower than 1 in the low ionic strength cases, and to an unknown degree. Current control therefore risks excessive Joule heating. To guard against that risk, one must stopper the top of all gels except the one at the highest ionic strength, turn on the current a 1 mA/tube and observe the voltage. Then one may unstopper the tubes and electrophorese at that level of the regulated voltage. Since the gel at the highest ionic strength gives rise to the maximal current at constant voltage it is the gel which generates the maximal Joule heat. Since it is known that it can be dissipated, this heat in all others can be dissipated even more so.

The same rationale for voltage control applies to the multichamber apparatus, since the various chambers are wired in sequence, not in parallel [41].

It should also be noted that the ionic strength cannot be increased by the addition to the system of amphoteric compounds with widely divergent pK s of the acidic and basic groups, such as amino sulfonates [79]. These compounds with pK s close to 0 and 14 do not affect the pH of the system and would appear applicable to raising the ionic strength of a buffer system. However, the charge interaction between amine and sulfonate prevents these compounds from providing conductance, and therefore also ionic strength.

9.3   REDOX Conditions in the Gel

Polyacrylamide gels, in view of their generation by a free radical catalyzed reaction, are highly oxidative. In a representative case, a 10 %T gel of 1.0 mL volume polymerized at pH 9.6 requires 1 μmol of an SH-reagent to produce net reducing conditions in the gel [106]. Although a number of reports in the literature have singled out persulfate as being responsible for oxidative damage of the protein during PAGE, the same mechanism of free-radical damage should prevail with other free-radical donors used to polymerize the gel, at least qualitatively. Quantitatively, the difference between using riboflavin and persulfate may be significant in this regard, since the former upon illumination gives rise to relatively few chain initiations and relatively large average chain lengths [51]. However, other considerations as well come into play which make it unlikely that persulfate is responsible for oxidative damage: It migrates out of the gel ahead of the protein; the rate of this migration can be monitored by benzidine staining of the gel [107]. And persulfate, although characterized by an exceedingly high standard potential thermodynamically, is also kinetically rather sluggish, requiring catalysis by such ions as Ag^+ [106]. A more likely mechanism for the oxidation of proteins migrating through gels independently of the particular kind of initiator used, whether charged and migrating or not, involves the secondary formation of uncharged peroxides in the gel. These cannot be removed under the conventional conditions of pre-electrophoresis, since the amino groups of the trailing constituents are not sufficiently reactive, except possibly at very high pH, to sweep out these peroxides. However, they can be swept from the gel by charged SH-compounds such as thioglycolate; under the conditions stated above, an excess (3 μ mol/mL of gel) of thioglycolate migrating ahead of the protein can provide net reducing conditions in a representative case.

Thus, whenever one suspects a loss of protein activity on the stacking gel due to the oxidative nature of polyacrylamide, the stacking gel analysis described in Chapter 7 should be repeated with 3, 5 and 10 μ mol of thioglycolate per mL of gel, neutralized to the trailing phase pH with the common constituent, and layered on top of the gel in a solution denser than the sample. The sample can then be layered over the thioglycolate solution. Upon electrophoresis, the thioglycolate pulls ahead of the protein and sweeps the gel free of peroxides. Another way of testing for REDOX effects would be to run stacking gels in agarose [97] or in Sephadex.

9.4 Solvent

9.4.1 Solubilization

A low recovery of activity in the stack may be due to the failure to solubilize the protein. Since it is being assumed that the proper solvent system for the protein has been investigated prior to electrophoresis in a stacking gel (Chapter 2), all we have to consider at this point is the likelihood that the suitable solvent has been modified or removed under the conditions of electrophoresis.

If the required solvent is a solution of charged detergent, or contains charged detergent, the electrophoretic removal of that detergent from the protein, or the removal of uncharged detergent by the charged one in form of a mixed micelle, are possible ways in which the protein may become progressively insoluble and therefore incapable of migrating within the stack. The first mechanism suggests a stacking gel analysis with various levels of charged detergent in the upper buffer, to find the level capable of compensating for the rate of electrophoretic removal of that detergent from the protein. The latter mechanism would require an analogous enriching of the upper buffer in the neutral detergent component until the rate of mixed micelle transport into the gel equals the rate of its removal from the region of the protein.

Gels run in 8 M urea may present a similar problem. The ion flux through the gel may, due to the hydration shells of the ions, dilute the urea in the gel progressively until the protein falls out. The remedy, again, is to prepare the upper buffer in 8 M urea.

Another type of effect which electrophoresis may exert on the solvent system needed to solubilize the protein, concerns the dependence of solubilization on the micellar state of a detergent [1],[3]. Thus, the Joule heat in the gel may greatly increase the micellar weight of non-ionic detergents, thereby modifying their solubilization characteristics. The ionic strength of the gel strongly influences the critical micelle concentration (CMC) of charged or amphoteric detergents. The CMC of SDS, for example, increases extremely sharply as the ionic strength is lowered below 0.1 M. This is of great consequence whenever solubilization depends on interaction of the protein with the micellar form of the detergent – it may simply not be available at the ionic strength prevailing in the stack, which due to the high protein concentration may be quite different from the predicted for the trailing phases in the buffer systems of Appendix 1 and 2.

9.4.2   Disaggregation

Stacked proteins have a tendency to aggregate due to the high protein concentration in the stack and due to the increased Joule heat in a zone of increased resistance due to the high protein concentration. Aggregated active proteins are frequently deficient or devoid of activities. The test and the cure for that form of inactivation is to run stacking gels made in various uncharged disaggregating media: High ionic strength, urea and detergents interfering with electrostatic, H-bond, and hydrophobic interactions, respectively (see Chapter 2).

The progressive admixture of charged detergent to uncharged detergent produces solvents which are increasingly capable of disaggregating the membrane protein, adenylate cyclase [4], although at the same time, such admixture proves to be progressively inactivating. Thus, to obtain the maximal degree of disaggregation compatible with maintenance of sufficient protein activity for detection, stacking gel analysis can be carried out as described in Chapter 7, except that each gel and upper buffer contains various amounts of charged detergent. Optimally, this experiment utilizes the multichamber tube apparatus, since different upper buffers are required [41].

9.5   Charged Cofactors such as Divalent Metals

The activity of a protein within the stack may have been lost through electrophoretic removal from the protein of a charged cofactor such as a divalent metal. Thus, stacking gels may be required with increasing levels of Mg^{++} or Ca^{++} in the lower buffer, in order to replenish these cofactors at the same rate at which they are lost from the protein through electrophoresis. The procedure for the stacking gel analysis is again that described in Chapter 7.

The near-quantitative recovery of the protein, or the protein activity, in the stack of optimal pH (Chapter 7), maximal trailing ion mobility (Chapter 8) and of suitable "happiness conditions" (Chapter 9) has not as yet resulted in any separation. Nonetheless, these results obtained on stacking gels alone, and by the analysis of a single gel slice to which the stack is confined, have solved the separation problem to such a degree that the remaining task of *separation* appears automatic and relatively trivial. All that needs to be done at this point to achieve separation is to unstack the stacked protein in a resolving gel of restricted pore size, thus exploiting molecular size differences between species for separation, in addition to their differences in relative hydrophobicity and net charge.

10 Physical Characterization of the Species of Interest and its Contaminants by Ferguson Plot

At the level of optimizing the pH by stacking gel analysis, the number of protein components in the sample was irrelevant. When we attempt now to characterize, identify and test for homogeneity the species of interest on the basis of migration distances at various gel concentrations, the number of proteins in the system remains irrelevant if the protein of interest can be detected by specific assay for an activity. But it becomes important, if that's not the case and if we are therefore forced to recognize the protein on the basis of its staining pattern. This is due to the fact that in the absence of a specific assay, there is no objective way to recognize which protein zone on a gel belongs to which protein, unless its concentration, or its stained color are unique among the many protein components forming stained zones on the gel. Quantitative PAGE applied to multicomponent systems is therefore limited to those fractionation problems in which the protein of interest can be recognized by assay or by a specific staining property. In all other cases, prefractionation by methods other than PAGE is needed to sufficiently simplify the system so that the protein of interest can be recognized in the gel pattern at different gel concentrations. Even the transverse pore gradient gel (Section 4.7.2) fails in this regard since %T at any one point along a pore gradient is insufficiently defined.

10.1 R_f as a Physical Constant

The stacked protein will now be introduced into a resolving gel characterized by a restrictive pore size and by an increased value of the trailing ion mobility. A restrictive pore size lowers the net mobility of the protein below that of the trailing ion, thereby unstacking it. Similarly, an increase in the trailing ion mobility in the resolving gel must, at one point, unstack the protein. Thus, unstacking of the protein can be brought about by either molecular sieving or a change of pH or both.

The unstacked protein at each gel concentration will be characterized by its R_f, defined as the ratio between the migration distance of the protein zone and the moving boundary in the resolving gel. If, as in many buffer systems of the Extensive Buffer Systems Output (Appendix 2), the leading ions in the stacking and resolving gels differ, two moving boundaries arise in the resolving gel (across Phases PI/LAMBDA and LAMBDA/GAMMA in the language of that output), with the leading ion of the stacking gel serving as the leading ion in the first case and as the trailing ion in the latter. In relation to the R_f measurement only the first of these boundaries, i.e. the one displaced directly ahead of the proteins is being used as a "front". Changing the trailing ion mobility of this moving boundary, allows one to select at will any desired value of R_f at a particular gel concentration. That is the prime usefulness of the various increasing trailing ion mobilities in column 2 of the Buffer Systems Tables (Appendix 1 and Fig.4).

The recognition of the stack in the resolving gel proceeds in the same fashion as the recognition of the stack in the stacking gel (Section 7.8). However, since the value of the trailing ion mobility is higher than that in the stacking gel, the likelihood of finding tracking dyes with mobilities between leading and trailing ions is much improved compared to the stacking gel.

In cases where the progressive introduction of stacked components (such as micellar SDS) widens the stack in proportion to the time of electrophoresis, an "extended stack" may result whose trailing edge recedes progressively. Under those circumstances, R_f values can only be referred to the leading edge of the extended stack, and care must be taken to ascertain that the dye used as a front marker migrates at this edge (see Chapter 15).

The R_f value is a physical constant characteristic of the protein at a controlled gel concentration and in a defined buffer milieu. All of quantitative PAGE rests on that proposition. However, its validity diminishes to the degree that the gel concentration and the buffer composition and temperature are uncontrolled and irreproducible. Thus, it is imperative to monitor pH and specific conductance of the stacking gel buffer and resolving gel buffer. By contrast, since the catholyte and anolyte are mere reservoirs for the trailing and common constituent, respectively, their control is not important.

It is more difficult to make reproducible gels than it is to define a buffer milieu. Close attention needs to be paid to all of the parameters that influence the polymerization reaction (Chapter 6). When this is done, the measured variance of R_f[108] and the variance of polymerization efficiency (% conversion) [74] can be reduced to values of a few percent.

10.2   Optimization of the Trailing Ion Mobility in the Resolving Gel

The variance of R_f can be reduced to an acceptable level when the values of R_f range between 0.25 and 0.85. Very low and very high R_f values are also irreproducible [108]. Thus, the trailing ion mobility must be chosen such that in the range of gel concentrations of interest, R_f values lie within the desired range. Practically, gel concentrations between 5 and 12 %T, 2 to 5 %C_{Bis}, are frequently the most convenient to polymerize and to use. One would therefore select an 8 %T gel to find a value of the trailing ion mobility such that the protein of interest exhibits an R_f of 0.55, i.e. intermediate between 0.25 and 0.85.

Trailing ion mobilities of increasing values are selected from column 2 of the Buffer Systems Tables (Appendix 1 and Fig.4), or from values obtained from those of column 2 by interpolation. The corresponding buffer recipes are given in columns 5–7 of the same tables. If interpolated values of the trailing ion mobility are used, the corresponding buffer recipes are most easily obtained by a graphic interpolation. The values given in columns 5-7 are plotted, and the desired values are read off the non-linear plots. It is not possible, of course, to predict which trailing ion mobility will yield an R_f of 0.55 in a gel of 8 %T. This has to be determined by a systematic experiment with resolving gels of 8 %T and systematically varied trailing ion mobilities.

When using the Extended Buffer Systems Output (Appendix 2), resolving gel buffers corresponding to the desired RM(1,PI) values (column 1 of the Subsystems Table of Fig.5 of Appendix 2) are selected from columns 11 and 12 of the same Subsystems Table. The final molar concentrations provided by these two columns for constituents 3 and 6 need to be multiplied by 4 to give the various resolving gel buffers for the subsystems. An easy mode of calculating these buffer compositions in g and mL of the constituents can be used as explained for the preparation of various subsystem stacking gel buffers in Section 8.1.

Resolving gels (1.2 mL/tube) are then polymerized in each of the subsystems with the various trailing ion mobilities. These are topped with 0.5 mL of the identical stacking gel (selected as described in Chapters 7–9). Electrophoresis is carried out, and the protein zone as well as the moving boundary front are detected by the procedures described for stacking gels (Sections 7.4 to 7.8). The subsystem buffer providing an R_f of 0.55 is selected or interpolated. It is designated by Roman numerals (see Section 7.1).

10.3   Ferguson Plot

The observation of a band, or the measurement of its R_f, provides no information concerning the properties of a native protein. Only when the molecular net charge of fully disaggregated and fully denatured protein subunits is made uniform by derivatization with SDS, can migration distance provide information concerning the molecular size. Within the framework of a general separation strategy for native proteins, the SDS-PAGE technique of estimating molecular size applies only to the determination of the subunit structure of native, post-translational proteins and will be discussed in that context (Chapter 15).

In contrast to the migration distance at any one gel concentration, there is information with regard to the properties of a protein in the linear semilogarithmic plot of the R_fs at various gel concentrations *vs* gel concentration. This plot of log R_f *vs* %T is designated as the Ferguson plot. The information to be derived from the Ferguson plot is the size and the net charge of the macromolecule, its degree of homogeneity and, most importantly for preparative purposes, the optimally resolving gel concentration for any pair of charged molecular species. A corollary to that capability of the Ferguson plot is that it can pinpoint a resolution optimum at "0 %T", i.e., in a non-gel, or "non-restrictive" gel. In that situation, the Ferguson plot directs the separation path into electrofocusing or preparative gel MBE.

The experimental task, in constructing the requisite Ferguson plot, is very simple. It consists of gel electrophoresis at 3 to 7 gel concentrations, at the known optimal pH and the known optimal trailing and leading ion mobilities in the stacking gel and the resolving gel. Using tube apparatus, it is quite easy to do that in duplicate in a single apparatus and experiment. The resulting R_f values for the protein of interest and the protein contaminants need to be measured. All subsequent operations leading to molecular weight, molecular net charge and optimally resolving gel concentrations are automatic, taking advantage of a series of computer programs developed by David Rodbard which are designated as the PAGE-PACK. The PAGE-PACK written in FORTRAN-IV is available from the Biomedical Computing Technology Information Center (BCTIC), Vanderbilt Medical Center, Nashville TE 37232, under program identification number "MED-34 PAGE-PACK".

10.3.1   Selection of Gel Concentrations

To construct a Ferguson plot, at least 3 gel concentrations are required. However, the 95% confidence envelopes computed for a line with 3 data points are enormous, making such a plot largely useless. On the other hand, these confidence envelopes do not improve anymore significantly, when more than 7 data points are used [108]. Therefore, 6 or 7 gel concentrations per Ferguson plot are sought in practice.

Procedurally it is best to select arbitrarily 3 widely divergent %T values first, to perform PAGE on those and to select the remaining 4 gel concentrations by interpolation. Although the intra-experimental variability is less than the inter-experimental one, and therefore a 2-stage electrophoretic experiment slightly increases the magnitude of the 95% confidence envelopes as compared to an analysis of all 7 gel concentrations in a single experiment, a 2-stage approach gives some experimental test for the adequacy of control over the conditions of polymerization and buffer composition. There is also an advantage of a 2-stage experiment with regard to the numbers of analyses in the construction of Ferguson plots for active proteins. Thus, the 8 %T gel used for optimization of the trailing ion mobility in the resolving gel (Section 10.2) or a single alternative gel must be analyzed fully, i.e. for a 1.2 mL gel tube alternate 40-50 gel slices must be analyzed. The 2 other widely divergent %T values tested at the first stage, e.g. a 5 and a 15 %T gel, may then be analyzed in the gel slices representing the bottom and top thirds of the gel only. The preliminary Ferguson plot based on 3 points then allows an even more economical assay of the 2nd stage of the experiment; usually only 3 to 5 gel slices can then be analyzed per

%T value. Thus, a Ferguson plot of an active protein can be constructed with about $25 + 2 \times 8 + 4 \times 4 = 57$ slice analyses, instead of the 175 analyses required if one were to fully analyze every gel.

10.3.2   Selection of Crosslinking Agent

The selection of the appropriate crosslinking agent and its concentration depends on the size of the molecular species under investigation, the range of desired gel concentrations and the analytical or preparative purpose of the fractionation.

For the construction of linear Ferguson plots, %C needs to be constant for all gel concentrations (%T) [70]. However, the biphasic relation between retardation in a gel (K_R) and %C (Chapter 5) allows one to select a %C value appropriate for small (5 %C_{Bis}), medium (2 or 3 %C_{Bis}) and big (10 to 30 %C_{Bis}) proteins. Presumably, the first two %C values yield long-fiber (1-D) gels with fiber radii of 1-2 nm, while the latter yields supercoils of 10-30 nm radius which at low %T approximate zero-fiber-length "point" (0-D) gels (see Chapters 5 and 6).

In practice, when the size range of the species of interest exceeds 100 000 in MW, the 15 %C_{DATD} crosslinked gel appear preferable [52],[109]. With smaller species the most useful gels appear those at 2 %C_{Bis}. These have sufficient strength of wall adherence to permit preparative fractionations with gels of 20 cm^2 of cross-sectional area, i.e. the largest preparative gels that can be adequately temperature controlled at practical rates of PAGE (Section 14.2). Since analytical and preparative PAGE have to be conducted under identical conditions to make data predictable and interconvertible, this argument in favor of 2 %C_{Bis} gels applies at the analytical scale. These gels, in contrast to the more restrictive 5 %C_{Bis} gels, are also fully transparent, which makes it easier to detect minor stained bands. Finally, they have the advantage of spreading out, relative to 5 %T gels, the range of gel concentrations which will yield R_f values between 0.25 and 0.85. This is an advantage because it requires less care in pipetting to make gels in intervals of 2 %T than in intervals of 0.5 %T; correspondingly, the experimental error is less, and the confidence limits of the Ferguson plot are tighter [78].

There are, however, analytical problems where one may wish to prepare at any %T the most restrictive possible gel, e.g. in application to very small molecules. This is certainly the 5 %C_{Bis} gel [73].

When working with very large species, in excess of 1 million in MW, the %C_{Bis} range from 10 to 50 %C would appear applicable, since pore size increases progressively in proportion to %C in that range (Fig.25). However, the adherence to glass walls and the mechanical stability of these highly crosslinked gels also decrease in proportion to %C_{Bis}. Moreover, polymerization mixtures above 30 %C_{Bis} become increasingly water insoluble and require the addition of solvents such as formamide [73]. Thus, open-pore media like the agarose-polyacrylamide "copolymer" (Section 5.4) or even pure agarose gels [40],[98] may be preferred. These, however, may raise problems with regard to electroendosmosis or to retardation and binding in "electroendosmosis-free" agarose [97] (see Chapter 6).

10.3.3   Procedure of Polymerization and Electrophoresis

The general procedure of polymerization and of electrophoresis described in Chapter 5 is followed. Resolving gels at the various gel concentrations are polymerized successively in the gel tube apparatus. Usually 1.2 mL/tube are used. After evacuation of one polymerization mixture, and the addition of TEMED, the polymerization lights are turned off temporarily while the polymerization mixture is being pipetted into the gel tube and overlayered with water. The upper buffer reservoir is then turned back to its original orientation to maintain vertical alignment and the illuminators are turned on again. After the last gel is polymerized for 30 minutes, the upper buffer reservoir is inverted, and the layering water is removed from each tube by tissue paper or cotton swab. The reservoir is replaced into the apparatus, returned to its original orientation, and 0.5 mL of the polymerization mixture for stacking gel are applied to each tube. After 30 minutes of photopolymerization, the overlayering water is removed as before, the reservoir is filled with upper buffer and placed into the lower buffer reservoir filled with lower buffer to the same level as the upper buffer. Sample is underlayered under the upper buffer and onto the gel surface of each gel, and PAGE is conducted at 1 to 2 mA/tube, 0–5 °C. When the tracking dye in one of the gels approaches the bottom of the tube within 2 mm, a rubber stopper is placed into the top of the tube and the tube is withdrawn from the bottom of the upper buffer reservoir. The stopper is firmly tightened in its grommet, and PAGE is continued. The gel is removed from the tube and either stained (Section 7.8.2.3) or sliced (Section 4.11.3) for assay.

10.3.4   Procedure of R_f Measurement

When the gel is evaluated by fixation and staining of protein bands (according to the procedures given in Section 7.8), the migration distance for each band and for the moving boundary front can be determined 1) by manual measurement under a magnifying glass or after photography; 2) by gel densitometry; 3) by electronic R_f-measuring device. In each case, the moving boundary is located by anyone of the methods discussed in Section 7.8.3.

When the gel is evaluated by gel slicing and measurement of activity, the R_f is determined for each band either 4) slice assay for protein activity, 5) by radioisotope analysis of transverse gel slices, or 6) by autoradiography of gels or longitudinal gel slices.

The R_f at each gel concentration is measured for the protein of interest as well as for the closest migrating contaminants which one is attempting to separate from the protein.

10.3.4.1 R_f Measurement by photography, caliper and ruler

The gel is photographed prior to its removal from the gel tube and as soon as possible after termination of electrophoresis, to avoid as much as possible diffusion spreading of the dye marking the moving boundary front. A Polaroid camera permanently set up in the laboratory above an illuminated plate and a transparent (Plexiglas) gel holder (Section 4.11.7) is most convenient for that purpose. The polarity of the gel is defined by inserting a fine hypodermic needle into the bottom of the gel. The needle may be dipped into India ink. The gel is then removed (Section 7.7) from the tube. At this stage, the position of the stack, if visible as a sharp refractory band or sharp band of tracking dye, can also be marked by needle and India ink. The gel is then fixed and stained by the procedure described in Section 7.8.4. The stained gel, contained in a storage tube marked at the very top with experiment and gel number, filled completely with fixative and stoppered by screw-cap, is re-photographed. The distance (x) between the origin of the resolving gel and the band is then measured on the photograph by caliper (Kern EKG-caliper No. 11121) and ruler, or on the gel aligned along a ruler, within the fraction of a mm. It is convenient to read the distance on the ruler under a magnification lamp (Fisher Scientific No. 12-071-1). The distance (y) between the origin and the marked stack position is measured in the same way. The R_f is calculated as $(x)/(y)$.

Alternatively to marking the stack position on the unstained gel, the ratio of migration distance of the stack ("front"), a, to gel length, b, may be measured. R_f is then calculated as $x \times b/a \times c$, where c = gel length of the stained gel.

If it is not possible to detect the stack position on the unstained gel visually, it must be calculated from the migration distance of a colored marker m, and the R_f of the marker at the particular gel concentration, $R_f(m)$, as $a = m/R_f(m)$.

10.3.4.2 R_f Measurement by densitometry

The stained gel is subjected to densitometry by any of a large number of densitometers for the visible range (e.g. Biomed No.SL-504-XL). The recording distances of the zone (x) and the moving boundary (y) are measured and divided to give the R_f as above.

10.3.4.3 R_f Measurement by electronic R_f-measuring device (cf. Section 4.11.3)

This is by far the least laborious and most accurate way of measuring R_fs of stained bands. The gel tube is placed into the groove of appropriate size on the instrument. The hairline is set onto the origin of the resolving gel and the potentiometer is set to 0.000 by the left dial. The hairline is then moved above the position of the front moving boundary, and the potentiometer is set to 1.000 with the right dial. Moving the hairline into any band position then provides a digital readout of the R_f value. This allows for rapid R_f measurements on consecutive bands by moving the hairline unidirectionally across the gel.

10.3.4.4 R_f Measurements by protein assay on transverse gel slices

The possibility to physically characterize the molecular species associated with some measurable function (bioactivity, immunoactivity, binding activity, radioactivity etc.) by PAGE rests on the determination of the R_f of the activity on suspensions of transverse gel slices. Depending on the mode of assay, the detection of the activity in the gel slice may require its elution from the slice, or the diffusion of a reagent into the slice. Depending on whether it is the former or the latter, the diffusion time required for substantial (70% or more) recovery of the activity may be prolonged or short; a protein may require anywhere from 3 to 48 h for diffusion from the gel slice, and its rate of appearance in the diffusate needs to be measured in each particular application.

After the appropriate diffusion time, the protein activity is measured in alternate resolving gel slices. The slice number of the peak activity is divided by the slice number corresponding to the moving boundary front to give its R_f. The slice number of the moving boundary front is determined by marking the tracking dye position with India ink, using a fine needle, prior to gel slicing. The slice marked by ink thus becomes recognizable as the position of the moving boundary front.

The possibility to measure R_f values on the basis of protein activity alone frees the physical characterization by PAGE of the need to isolate the activity. Since such isolation always brings about a certain degree of loss of native properties, the possibility to characterize by PAGE from crude solution is beneficial from the viewpoint of activity and conformation maintenance. A further advantage arises with regard to load economy: Activity detection from crude solution is only limited by the sensitivity of activity detection. The requirement for microgram amounts of protein per cm^2 of gel, to saturate the adsorption sites of polyacrylamide [50], is satisfied under those conditions by the contaminant proteins. The sensitivity of detection of protein activities is, of course, usually much higher than that of protein staining.

10.3.4.5 R_f Measurement by isotope analysis

All that has been said in the preceding section applies to the analysis of radioactive proteins on transverse gel slices. Only a small number of specific features of R_f determination of radioactive proteins need to be pointed out: First, the gel may quench the radioactivity. In those cases, one may choose to apply DATD crosslinked gels and dissolve them by periodate; or one may dissolve Bis-crosslinked gels in hydrogen peroxide at 50 °C. Secondly, the position of the front may be recognizable by the free label if it is charged. Thirdly, the label may partition quantitatively into the counting fluid (e.g. toluene) if it is hydrophobic and non-covalently attached to the protein. In that case (e.g. steroid binding proteins) the recovery of label in the counting fluid is nearly instantaneous. Finally, R_fs may be determined on autoradiograms as described in Section 10.3.4.1, following autoradiography (Section 4.11.9) of gels or longitudinal gel slices (see Section 4.11.4).

10.3.5 Procedure of Computing the Ferguson Plot and Evaluation of the PAGE-PACK Output

The R_f values at the various gel concentrations are entered into datafile FILE01 of the PAGE-PACK (Fig.9). Using these data, the Ferguson plot and its 95% confidence envelopes are computed by running program RFT1 of the PAGE-PACK. The 4-page output (Figs.10 to 12) states on page 1 (Fig.10) the values of K_R and Y_o, and for a weighted regression, the correlation coefficient R for the best-fit-line, and the standard deviations of K_R (here designated as D) and Y_o. On page 2 of the output of RFT1 (Fig.11), the joint 95% confidence envelopes of K_R and Y_o are given in numerical terms. When Y_o values are plotted along a logarithmic ordinate, and the K_R values (here designated as *slope*) along a linear abscissa, using semi-log graph paper, these confidence envelopes appear as ellipsoids. Page 3 of the RFT1 output (Fig.12) graphs the Ferguson plot, with the inner 95% confidence envelopes around the line and the outer envelopes applicable to a single observation along the line. Page 4 (not shown) plots the same data on a linear scale.

In order to take advantage of the later computation of the optimally resolving pore sizes for separating the species of interest from those migrating just in front of and just behind it, the Ferguson plots of these species are determined in the same fashion as that for the unknown protein.

It is evident from Figs.9 to 12 that the main benefit to the non-statistician of the PAGE-PACK over a manual graphing of the Ferguson plot consists of the error-free calculation of the best-fit-line, its slope and y-intercept, and of the statistical treatment of the data. Clearly, the entering of the input data and the computation takes only a few minutes, particularly once a terminal is available in the laboratory. The outward complexity of the output with its myriad of symbols and numbers is of no practical consequence: Only those very few parameters discussed here are relevant in the routine use of the PAGE-PACK.

10.4 Molecular Characterization and Homogeneity Testing on the Basis of K_R and Y_o

10.4.1 Identity

A molecule in a particular buffer and solvent milieu is defined by its size (K_R) and net charge (Y_o). This set of parameters can by plotted as a point on a graph of Y_o vs K_R. When the joint, i.e. interdependent, 95% confidence limits of the 2 parameters are plotted, to represent the range of probable positions of this point, an ellipsoid area results. These areas are designated as the joint 95% confidence envelopes of K_R and Y_o; they are computed as described in Section 10.3.5. In short, we will call them "K_R-Y_o ellipses". A molecular species in a particular solvent and buffer milieu is physically defined by its K_R-Y_o ellipse.

Between-experimental identity or non-identity of 2 species can be determined by ellipse criteria as follows. Species exhibiting overlapping ellipses are indistinguishable with 95% confidence. Species exhibiting non-overlapping ellipses are distinct. Species exhibiting partially overlapping ellipses are distinguishable at a confidence

Input File "MUNI"

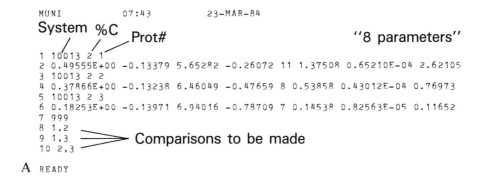

```
MUNI              07:43            23-MAR-84
System %C   Prot#                           "8 parameters"
1 10013 2 1
2 0.49555E+00 -0.13379 5.65282 -0.26072 11 1.37508 0.65210E-04 2.62105
3 10013 2 2
4 0.37866E+00 -0.13238 6.46049 -0.47659 8 0.53858 0.43012E-04 0.76973
5 10013 2 3
6 0.18253E+00 -0.13971 6.94016 -0.78709 7 0.14538 0.82563E-05 0.11652
7 999
8 1,2
9 1,3     ======> Comparisons to be made
10 2,3
```

A READY

Program IDENT

Output 1

```
CURVE   PROTEIN D.F.    RES.VAR.
  1        1        9   6.52100E-5
  2        2        6   4.30120E-5
  3        3        5   8.25630E-6
G=   8.86244E-4  E =   20 MEAN RES.VAR. = G/E = 4.43122E-5
 B =  5.10291   C =  1.0713   CHI SQUARE  =B/C =  4.76331
COMPARE WITH CHI-SQUARE DISTRIBUTION WITH  2  D.F.
APPROXIMATE CHISQUARE WITH 2 D.F. IS 5.93538 FOR P=0.05

ENTER YES IF YOU WANT TO USE COMBINED RES. VAR. AND D.F.
ENTER NO  IF YOU WANT TO ANALYSE EACH CURVE SEPARATELY
  ?yes
DO YOU WANT  JOINT C.L. ELLIPSES, YES OR NO ?
  ?yes
```

B

Figure 26. Representative input file for PAGE-PACK program IDENT. (A) The "8 parameters" derive from the output of program RFT1 (lines 5 and 6, counted from the top, of Fig.11). (B) Top lines of the output of PAGE-PACK program IDENT providing both the data base for and the mode of response for the decision whether or not the data are statistically homogeneous, thus allowing for computation of "reduced 95% confidence ellipses".

level that decreases with the degree of overlap. Thus, in these questionable cases of ellipse overlap, a re-analysis of the Ferguson plots in a single experiment is needed to provide a statistically homogeneous sample of data points (R_fs). A statistically homogeneous set of R_fs allows one to assume an identity of error sources in the analysis of the 2 proteins; this assumption allows for a shrinking of the K_R-Y_o ellipses. Ellipse shrinking is carried out by means of computer program IDENT (written in BASIC for the DEC-10 computer, by D. Rodbard, NIH. A listing of that program is given in Appendix 5). The input into this program consists of the 7 parameters printed out on lines 5 and 6 of Fig.11 for each protein (lines 2 and 3, Fig.26A). After entering these parameters for each of the components to be compared into a data file, one specifies which component is to be tested for identity with which other (lines 8–10, Fig.26A). Program IDENT then requires a decision as to the statistical homogeneity of the sample. If all data to be compared derive from a single experiment or at least from consecutive runs with identical solutions and under otherwise the same conditions, the statistical sample of R_f values may be considered homogeneous, and the use of "combined residual variances and degrees of freedom", and of "joint confidence limit-ellipses" is justified ("yes" answers in lines 11 and 13, Fig.26B). Program IDENT then computes the reduced ellipses (Fig.27A). The output format is that described for Fig.11. Program IDENT then tests the identity of the species independently on the basis of F-test criteria [108] (Fig.27B). If the F-ratio is greater than the number stated in the output (line 3 for each protein in Fig.31B), the two protein components in question are distinct by F-test criteria.

It is important in the identity testing by ellipse criteria that the sizes of the ellipses which are being compared are similar. If one of the 2 ellipses is very large, indicating very poor fit or paucity of the R_fs constituting the Ferguson plot, analysis should be repeated with more data points until ellipse sizes become comparable.

It should also be realized that the R_f on a single gel may provide a not only experimentally more easily obtainable but a more definitive measure of distinction than the non-overlap of a K_R-Y_o ellipses [28], since it avoids between-experiment variability of migration distances. However, absolute values of migration distance of identical proteins may vary on a simple gel as a function of the volume, conductivity and viscosity of the sample. Also, when using identical migration distances as evidence for identity, it needs to be kept in mind that mobilities of 2 different proteins are identical at the μ-point (Fig.28), and that for a determination of the optimally resolving gel concentration a Ferguson plot is required (Chapter 11).

10.4.2   Homogeneity

Commonly, gel electrophoretic fractionation aims at obtaining a gel pattern consisting of a single band. However, single-bandedness is both an insufficient criterion of homogeneity as it is unrealistically ambitious. It is insufficient because the particular gel concentration and/or pH used may not be conducive to separation. Thus, it will be shown further below that many protein pairs exhibit Ferguson plots which

Output 2

```
CURVE 1 SYSTEM   10013   C=   2   PROTEIN   1

SLOPE            YO-LOWER        YO-UPPER

% SQRT OF NEGATIVE NUMBER IN LINE 560
-0.144657        3.60576         3.60585
-0.139224        3.26057         3.46163
-0.13379         3.0239          3.24022
-0.128356        2.83049         3.00503


----------------------------------------------------------------


CURVE 2 SYSTEM   10013   C=   2   PROTEIN   2

SLOPE            YO-LOWER        YO-UPPER

% SQRT OF NEGATIVE NUMBER IN LINE 560
-0.152434        3.22255         3.22265
-0.142407        2.64646         2.91199
-0.13238         2.26295         2.52711
-0.122353        1.96385         2.16089


----------------------------------------------------------------


CURVE 3 SYSTEM   10013   C=   2   PROTEIN   3

SLOPE            YO-LOWER        YO-UPPER

% SQRT OF NEGATIVE NUMBER IN LINE 560
-0.191252        3.46916         3.46929
-0.165481        2.09611         2.51964
-0.13971         1.36891         1.69303
-0.113939        0.919818        1.10567
-8.81676E-2      0.668034        0.66806
```

A

Output 3

```
CONTRAST OF CURVES 1  AND  2
COMBINED RESIDUAL VARIANCE =  4.43122E-5  ON  40 DEGREES OF FREEDOM
FOR SIGNIFICANCE , F MUST BE GREATER THAN  3.23173  ◄────────
  F-RATIO ON 40  D.F. IS  47.4524  ◄────────

CONTRAST OF CURVES 1  AND  3
COMBINED RESIDUAL VARIANCE =  4.43122E-5  ON  40 DEGREES OF FREEDOM
FOR SIGNIFICANCE , F MUST BE GREATER THAN  3.23173
  F-RATIO ON 40  D.F. IS  172.039

CONTRAST OF CURVES 2  AND  3
COMBINED RESIDUAL VARIANCE =  4.43122E-5  ON  40 DEGREES OF FREEDOM
FOR SIGNIFICANCE , F MUST BE GREATER THAN  3.23173
  F-RATIO ON 40  D.F. IS  76.3086
```

B

Figure 27. Representative output page of PAGE-PACK program IDENT providing (A) "reduced 95% confidence ellipses" and (B) identity testing between each selected set of two species on the basis of F-test criteria.

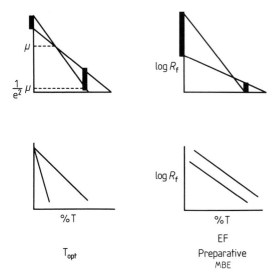

Figure 28. Dependence of the choice of an electrophoretic separation method on the particular separation problem revealed by the Ferguson plot. If size (K_R) differences among the proteins to be separated predominate or exist exclusively, PAGE at the T_{opt} is indicated (left panels). If charge (Y_o) differences are predominant or exist exclusively, separation by MBE or EF is indicated (right panels).

cross one another; at the crosspoint (μ-point, Fig.28) separation is nil. Moreover, ironically, such crosspoints frequently occur near 7 %T which is probably the most commonly used gel concentration.

The criterion of single-bandedness is also unrealistically ambitious since most protein activities are associated with multiple forms of the protein differing either in size (size isomers) or in net charge (charge isomers). This is due to concentration dependent aggregation which most proteins, if not all, undergo. And it is due to the loss or gain of charged ligands. Examples for charge gain are the binding of fatty acids [110], formate [105], molybdate [111], phosphate or nucleotides. Charge loss may be due to desialylation [56], [67], deamidation, or partial proteolysis which at neutral pH confers a partial negative charge on the protein due to the full dissociation of -COOH, at a fractional protonation of -NH_2 [112]. Some of these forms may be artifacts of isolation, others may be biologically significant (e.g. partially hydrolyzed, bioactivated human growth hormone [112]). Irrespective of their significance, the various size or charge isomeric forms of most proteins give the appearance of heterogeneity when one applies the criterion of single-bandedness to gel patterns, while in a more meaningful sense, homogeneity may have been already achieved, i.e. the various bands may constitute a charge and/or size isomeric family of the identical protein.

This situation is clearly evident in inspection on the basis of K_R-Y_o ellipse criteria. Size isomeric ellipses are displaced horizontally from another along the K_R axis, sharing the same Y_o. Charge isomeric ellipses are displaced vertically from another along the (logarithmic) Y_o axis, sharing the identical K_R. Thus, upon Ferguson plot

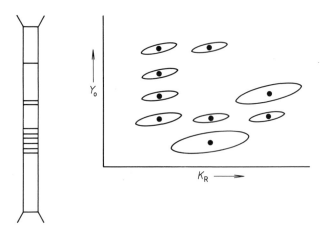

Figure 29. Molecular homogeneity testing by "ellipse criteria". A count of bands would provide a misleading impression of heterogeneity, while the plot of Y_o *vs* K_R reveals on inspection that all 9 components shown are either charge isomers (vertically aligned) or size isomers (horizontally aligned) of the same species, indicating homogeneity in a more meaningful sense.

analysis, and plot of the K_R-Y_o ellipses as described in Section 10.3, a family of ellipses displaced from another either horizontally or vertically presents a realistic piece of evidence for molecular homogeneity under the conditions of PAGE used while the appearance of component multiplicity in the gel pattern at a single gel concentration is misleading (Fig.29).

10.5   Physical Characterization by Molecular Radius and Weight

The physical characterization of the protein of interest in terms of its molecular radius and weight has limited intrinsic value since it does not provide any information in addition to that which the K_R already provides. The "translated" parameters, molecular radius and weight, are useful mainly as a basis of comparison with molecular size estimates derived from methods other than gel electrophoresis. However, it is important to realize that the K_R is an assumption-free experimental parameter descriptive of molecular size, while the translation to the conventional parameters of radius and weight is subject to a number of assumptions which render the translated parameters less accurate than the K_R value from which they are derived.

The first such assumption is that of either spherical or random-coil conformational type. No real protein is, of course, purely spherical or random-coiled. In reality, the full spectrum of intermediate forms between these models exists, and other models, such as "rods", also make a contribution to conformation which is being neglected here. In practice, any denatured protein, or very asymmetric protein, or flexible ribonucleic acid, lipopolysaccharide or charged carbohydrate, will be classed as "random-coil", any native protein as "spherical".

The second assumption is that of a common partial specific volume ($\overline{V}$) between the protein of interest and the standards. Usually, we will assume a $\overline{V}$ of 0.74 for unconjugated proteins, a lesser value (e.g., 0.66) for glycoproteins and a higher value for lipoproteins. Again, no matter which value is assumed, $\overline{V}$ will vary in reality among proteins. It is also difficult to locate molecular weight standards which fall into such classes as "glycoproteins" or "membrane proteins" (glycolipoproteins). Yet, to the degree that one fails to do so, the validity of the translation diminishes.

The third assumption consists of the neglect of protein hydration. Yet in reality, hydration has a great impact on molecular size. For instance, the β-subunit of the glycoprotein, human chorionic gonadotropin, exhibits K_R values very similar to the entire protein molecule, presumably because it carries the bulk of the carbohydrate moiety which is highly hydrated [67].

The error introduced into the translation of K_R by making these 3 assumptions is the same for the calculation of molecular radius and weight. To calculate molecular radii of standard "spherical" proteins, one assumes sphericity of these proteins; their volume, $4/3\pi r^3$, multiplied by $\overline{V}$ (mL/g), is set equal to their mass to calculate the molecular radius, R (nm). Then these calculated R values are used to derive the molecular radius of the unknown (see Section 10.5.2). In the calculation of molecular weight of a "spherical" protein, the volume of the sphere is derived from the radius, multiplied by $\overline{V}$, to arrive at a calculated mass by an exact reversal of those operations.

Error in the molecular weight of the standards also contributes to the error in the values of molecular radius and weight from K_R. Frequently, the molecular weights of the standards have been obtained by gel filtration and sedimentation velocity (isopycnic sedimentation). In those cases, it is important to realize that neither gel electrophoresis, nor any other "molecular weight method" except sedimentation equilibrium, measures molecular weight (mass, daltons) [113]. All other methods are responsive to molecular geometry (size, surface area, nm). It is not possible to arrive at meaningful molecular weight data by the methods responsive to geometry because of the uncertainty that they measure Stokes' radii; because of the compounded error from the relatively large confidence limits of Stokes' radius, $\overline{V}$ and sedimentation coefficient; and unavailability of homogeneous standards [113].

The type of molecular radius referred to in this treatment as $\overline{R}$ deserves a short comment. $\overline{R}$ here designates the geometric mean radius, or the radius of an "equivalent sphere". That means that physically, no sphericity is assumed. Rather do we express the size of a randomly shaped molecule as that of a sphere with the same surface area (i.e. an "equivalent sphere"). This avoids making unsubstantiated assumptions concerning the kind of molecular radius to which gel fractionation, or other fractionation methods, are responsive (differences in the values of the radii provided by various methods for the same molecule are described on page 123 of [16]). The common assumption, e.g., that gel filtration or gel electrophoresis "sees" the Stokes' radius is not experimentally verified and possibly not verifiable by presently available methods [113].

10.5.1   Selection of Molecular Weight Standards

As an example, let us consider the selection as molecular weight standards of "spherical" native unconjugated proteins. At least 7 such proteins varying in molecular weight and belonging either to the relatively acidic or basic proteins are required. These are tested for their ability to migrate within the stack of the buffer system optimized for the protein of interest (Chapter 7). Since all subsystems of stacking gels are compatible with a single resolving gel of the same buffer system, it is not necessary that the standard protein stack in the same subsystem as the protein of interest. Often, one may apply an stacking gel buffer to the standards which has a lower trailing ion mobility than the unknown protein. Once a standard is stacked, a single resolving gel is run to determine whether the preparation of the standard is homogeneous enough to allow one to discern a major band; it is then assumed that the major band comprises "the protein". If that is not the case, as in many commercial protein preparations, the search for a more homogeneous preparation must be repeated, until a preparation with at least a single major protein component is found. It is easiest, to conduct such a search for suitable standard proteins on a vertical gel slab. To improve the confidence limits of molecular radius and weight, the selection of as many as 20 standards is very desirable. This improvement in confidence limits is due to the fact that such a large number of protein standards tends to randomize the error due to the assumptions of spherical configuration, value of $\overline{V}$ and zero hydration. Fortunately, the considerable effort involved has to be spent only once for each buffer system.

Commercially available standard kits are not too useful in this regard at this time, since they contain proteins of a wide variety of charge densities and since acidic and basic proteins are usually not segregated. While this does not necessarily affect their usefulness in SDS-PAGE (see Chapter 15), it makes these kits frequently inapplicable to native proteins.

A very promising approach to making protein standards would consist of cross-linking of small molecular weight proteins of the various types, with enrichment of the higher homologues in a standard preparation by use of preparative methods. At this time, however, such homologous series of proteins of the various charge-types and conjugation-types are not commercially available, although the principle has been applied to the production of SDS-PAGE standards (e.g. BDH Biochemicals, Poole, UK, No.44223, or the gelatin oligomers produced by Serva, Cat. No. 21250-21254).

10.5.2   Determination of the K_R Values for the Standards

To be able to derive the molecular radius of the unknown from a comparison of its K_R with the K_R values of the molecular weight standards, one needs to determine these K_Rs, i.e. the Ferguson plots of the standards. The procedure is the same as described for the unknown in Sections 10.3.3 to 10.3.5. As a shortcut to running 7 to 20 Ferguson plots with 7 points each, an attempt should be made to utilize the same gel concentrations for the proteins of various sizes as much as possible. Thus, one may run all the proteins on 6 vertical gel slabs (2 slabs per apparatus, see Section 4.3) in 2 or 3 experiments, and determine all the K_R values within 3 workdays.

10.5.3 Procedure of Computing from K_R the Translated Parameters Descriptive of Molecular Size, i.e. Molecular Radius and Weight

On line 3 of the input file DATA03 (Fig.13), the term KR is entered to specify the type of function on which MW computation by program GIANTRUN is to be carried out, i.e. $(K_R)^{1/2}$ *vs* $\overline{R}$ and K_R *vs* MW.

The standard proteins, each defined by a protein number in data file RADII4 (Fig.30), and their K_R values are entered into data file DATA03 (Fig.13). If the standard proteins are not listed in RADII4, they may also be given an available protein number (of 999 or less) and be entered into file DATA03 together with their molecular radii and weights. The radius is calculated as

$$r = [(MW \times \overline{V}) \times 10^{-5} \times 3/(\pi \times 4)]^{1/3},$$

assuming molecular sphericity (see above). For convenience, that calculation may be made by computer. A suitable program in BASIC for the DEC-10 computer (program RADIUS of D. Rodbard, NIH) is listed in Appendix 6.

Program GIANTRUN computes the molecular radius and weight of spherical species by use of the linear relation of $(K_R)^{1/2}$ *vs* $\overline{R}$ (pages 2 and 3 of the output). It also computes the molecular radius and weight for random-coiled species by use of the linear function of K_R *vs* MW (pages 4 and 5 of the output). Page 1 of the output of GIANTRUN recapitulates the input data (Fig.13). Page 2 of the output (Fig.14) lists the molecular radii and weights of the unknowns, together with their 95% confidence limits. It also lists the correlation coefficient of the unweighted regression of $(K_R)^{1/2}$ *vs* $\overline{R}$. Page 3 of the output (Fig. 15) plots $(K_R)^{1/2}$ *vs* $\overline{R}$ with the 95% confidence envelopes around the line (inner brackets) and for a single observation around the line (outer brackets). Pages 4 and 5 then repeat the treatment of the data for the random-coil case, using the same format as that shown for pages 2 and 3 (output format not shown).

Again, as in the computation of the Ferguson plot (Section 10.3.5), only the relatively few parameters discussed here matter in practice. The error-free calculation of molecular radius and weight on the basis of the best-fit linear plots for spherical or random-coiled species, and the calculation of 95% confidence envelopes for these lines, are the main benefits of computation over a manual graphic approach. It should be noted in particular, that rigorous 95% confidence limits of molecular weight are only rarely found among literature reports, primarily because biochemists are only rarely self-reliant statisticians, and rarely collaborate with professional statisticians on those matters. The use of the PAGE-PACK obviates these problems and avoids the ambiguities of reporting standard deviations of molecular weight that refer only to the single data point, not the standard curve, and that are uninterpretable when the number of measurements is not reported.

METHYLGREEN	458.	.5122	15.
BROM-O-BLUE	670.	.5814	14.
ANGIOTENSIN-I	1297.	.7246	83.
BACITRACIN	1411.	.7452	67.
I 131ANGIOTENSIN	1421.	.7470	84.
INSULIN-B-CHAIN	3400.	.9991	71.
GLUCAGON	3485.	1.0073	70.
ACTH	3500.	1.0088	91.
INSULINDIMER	11466.	1.4982	64.
CYTOCHROME-C	12400.	1.5379	1.
RIBONUCLEASE-A	12700.	1.5502	2.
A-LACTALBUMIN	14437.	1.6178	40.
LYSOZYME	14500.	1.6202	3.
MYOGLOBIN	17800.	1.7348	4.
B-LACTOGLOB-SUB	18400.	1.7541	41.
PAPAIN	20700.	1.8243	51.
HGH	21500.	1.8475	5.
OVINEPROLACTIN	22550.	1.8771	21.
SOYBEANTRYPSINI	22700.	1.8813	50.
INSULINTETRAMER	22932.	1.8877	65.
CHYMOTRYPSIN	25100.	1.9454	6.
CHYMOTRYP-OGEN	25100.	1.9454	24.
CARBONICANHYDRA	31000.	2.0872	56.
CARBOXYPEPTID-B	34300.	2.1588	85.
BLACTOGLOBULIN	35000.	2.1734	54.
PEPSIN	35500.	2.1837	7.
LIPASE-PANCREAS	37800.	2.2298	86.
FSH	38400.	2.2416	32.
PEPSINOGEN	40400.	2.2798	63.
OVALBUMIN	43500.	2.3367	8.
ACIDGLYCOPROT	44100.	2.3474	61.
A-AMYLASE	45000.	2.3633	88.
FETUIN	48400.	2.4214	19.
A-1-ANTITRYPSIN	55000.	2.5268	22.
TRANSCORTIN	58500.	2.5793	36.
HEMOGLOBIN	64500.	2.6646	9.
BSA-MONOMER	67000.	2.6986	10.
PEPSIN-DIMER	71000.	2.7512	53.
TRANSFERRIN	74000.	2.7895	57.
PEPSNOGEN-DIMER	80800.	2.8724	163.
ENOLASE	82000.	2.8866	35.
ALCDEHYDRYEAST	141000.	3.4558	89.
OVALBUMIN-DIMER	87000.	2.9441	66.
LACTOPEROXASE	93000.	3.0103	45.
PHOSPHORYLASEB	185000.	3.7859	90.
GLUC6PHOSPHDH	104000.	3.1246	69.
BSA-DIMER	134000.	3.4000	16.
LACTICDEHYDR	140000.	3.4500	68.
ALDOLASE	161000.	3.6145	87.
GAMMA-GLOBULIN	160000.	3.6070	11.
CERULOPLASMIN	160000.	3.6070	48.
CATALASE	232000.	4.0826	60.
ACETYLCHOLINEST	240000.	4.1290	55.
PHYCOCYANIN	266000.	4.2730	37.
PHYCOERYTHRIN	290000.	4.3979	18.
FIBRINOGEN	340000.	4.6374	12.
PHOSPHORYLASEA	370000.	4.7699	47.
FERRITIN	450000.	5.0915	44.
APOFERRITIN	450000.	5.0915	52.
BGALACTOSIDASE	520000.	5.3429	62.
GLUTAMINESYNTH	600000.	5.6039	93.
THYROGLOBULIN	669000.	5.8139	13.

Figure 30. Representative list of protein molecular weight standards in the PAGE-PACK datafile RA-DII4. Geometric mean radii are computed assuming sphericity of the proteins, using program RADIUS (Appendix 6). The column on the right lists the arbitrarily assigned identifying numbers for each of the standard proteins.

The computation of molecular radius is applied to the K_R values of the contaminating species migrating just ahead and just behind the protein of interest, to allow for the subsequent computation of the optimally resolving gel concentration for the separation of these species from the protein of interest.

11 Pore Size Optimization

In addition to the characterization of the species of interest in terms of its size (K_R) and net charge (Y_o), the computation of the optimal gel concentration for the separation of the species of interest from its most closely migrating contaminants is the prime purpose and benefit of the PAGE-PACK to practical separation.

Since the contaminating species have to be identified by protein staining, the restriction of Ferguson plot analysis to oligocomponent systems in case of zone detection by staining (see introduction to Chapter 10) applies in the case of pore size optimization. This restriction is even more stringent here, since we are not only interested in associating particular bands with particular proteins at various gel concentrations, but we attempt to define which protein it is that migrates just in front of, or just behind the protein of interest at a given gel concentration. When the Ferguson plots of proteins intersect, the protein which is the closest migrating species at one gel concentration may be far removed at another and *vice versa*. Therefore, even in case of relatively few components, one may be forced into a preparative technique, i.e. gel excision in the region of the closest contaminants and Ferguson plot on the gel slices, to arrive at the K_R and Y_o of the closest migrating contaminants. Such 2-stage fractionation would require 18 mm diameter gels and 10-fold load at the first stage. The gel slices of interest are subdivided and the sections are directly re-electrophoresed under the conditions of the first stage, fixed and stained.

For the purpose of explaining the procedure of pore size optimization we will assume that it has been possible to characterize by K_R and Y_o the protein of interest as well as the contaminant species migrating in front of, and behind it which one wishes to separate it from. For each of the 3 species, $\overline{R}$ is determined by means of a standard curve as described in Chapter 10. The 3 parameters K_R, Y_o and $\overline{R}$ for the protein of interest (designated here as species A) and one of its 2 closest migrating contaminants (here designated as species B) are entered into data file DATA06, together with the temperature and the approximate duration (in minutes) of the run (Fig.17). Program TOPT is then run. The 1st page of the output (Fig.18) states the gel concentration of optimal resolution between A and B (T_{opt}), taking their bandwidths into account (assuming time dependent diffusion spreading according to Stokes' law). It also states the gel concentration providing the maximal separation between the peak positions of A and B (T_{max}). Page 2 of the output of program TOPT provides a plot of resolution (designated in the plot by R) and of separation (designated in the plot by S) as a function of gel concentration (Fig.19). Whenever resolution and separation are maximal at 0 %T on that plot, page 1 of the output also explicitly advises the user to employ isoelectric focusing (Chapter 12) or preparative MBE (Chapter 13) for resolving species A from B.

The procedure is then repeated with the second closest migrating contaminant as species B. Necessarily, the values of T_{opt} and T_{max} will turn out to be different, forcing one to either employ a 2-stage fractionation at each of the 2 suggested pore size optima successively, or a fractionation at a compromise-gel-concentration between the two optima with regard to the two contaminant species.

In general, the computation of the optimally resolving gel concentration between A and B will direct the fractionation path on the objective basis of the Ferguson plot analyses (Fig.28) into either an analytical and preparative PAGE fractionation (Chapter 14) or into an analytical and preparative charge fractionation, i.e. electro-focusing (Chapter 12) or preparative moving boundary electrophoresis (Chapter 13). PAGE fractionation is indicated whenever A and B differ predominantly or solely in K_R (Fig. 28, left). Charge fractionation is indicated whenever A and B differ predominantly or solely in Y_o (Fig. 28, right).

12   Electrofocusing on Gels

Let us assume that the computation of an optimally resolving pore size (Chapter 11) has resulted in a T_{opt} value of 0. This result directs us into one of the charge fractionation methods. At the analytical scale, zone electrophoresis in the absence of molecular sieving effects (agarose gel electrophoresis, cellulose acetate electrophoresis), ion exchange chromatography and electrofocusing on "non-restrictive" gels offer themselves for that purpose. Of these methods, chromatofocusing [114] and electrofocusing have the advantage that they automatically operate at a pH where charge fractionation is most effective, i.e. at or near the pI. The former method is preponderantly a preparative one, while EF lends itself to analytical separations at the microgram scale and to a high multiplicity of analyses. It is, however, much more than zone electrophoresis at an optimal pH, burdened by specific problems which will be discussed below.

In view of the possibility in any separation problem that the optimization of pore size directs one into electrofocusing (Fig.28), this separation method has been included under the heading of "Quantitative PAGE". However, there are additional reasons as well for considering EF not as a separate, independent separation method but as an integral part of one electrophoretic science. These are: a) The fact that the same theoretical treatment of moving boundary electrophoresis is capable of predicting the pH range and dynamics of pH gradients [20],[115]; b) the fact, that non-amphoteric acids and bases of differing pK can generate natural pH gradients [115],[116] and that therefore electrofocusing is not necessarily isoelectric, i.e. alignment of constituents is not in order of pI, but in order of net mobility, as it is within a system of sequential moving boundaries (a stack). We have therefore replaced the common term, "isoelectric focusing" (IEF) by "electrofocusing" (EF). However, in practice EF can hardly be considered a "quantitative" method, since at present the prediction of the properties of pH gradient systems is limited to buffer constituents with known ionic mobilities, a very short list indeed (see [195]; Fig.1 of Appendix 2). Obviously, this impediment will fall once these measurements are taken for larger numbers of suitable buffer species.

Electrofocusing was originally carried out on sucrose density gradients rather than on gels. But when applied to sucrose density gradients, the method has been a preparative one, since microtechniques for forming density gradients are not commonly available which would allow for its application to a microgram-analytical scale comparable to that of PAGE. Furthermore, the sucrose density method (and probably also chromatofocusing) suffers from pertubation by the frequent insolubility of proteins at or near their pIs, at least at protein loads commensurate with the column dimensions used. For these reasons, we will limit ourselves here to isoelectric focusing carried out on gels.

12.1   Apparatus

We will limit ourselves to those apparatus designs and procedures which a) allow one easily to follow the pH of a zone as a function of electrofocusing time, and b) provide efficient Joule heat dissipation as well as uniformity of temperature across the gel matrix. At this time, as will be explained in the subsequent sections, the same gel tube apparatus used for PAGE is preferred for both reasons. This preference based on scientific reasons runs counter, at present, to the preference of instrument manufacturers for electrofocusing on horizontal slab apparatus which is based on the ease of construction of such apparatus and on the commercial prevalence of the clinical market over the scientific one, with its obvious interest in pattern comparison under fixed conditions of electrofocusing.

Electrofocusing is a steady-state separation method. It is only at the "isoelectric end point" characteristic for a given protein that the pH of the zone of that protein becomes constant, i.e. the apparent pI (Fig.31), that all aggregation states of a given protein coalesce (Fig 1. of [64])and that proteins uniformly loaded within the gel combine into a single zone. Furthermore, an electrofocusing pattern becomes reproducible in its "Gestalt" only when all components of the pattern have reached their "isoelectric end points". That fact even holds for the case of a perfectly stabilized pH gradient which under appropriate conditions is theoretically possible [20],[115] and which experimentally may be achieved through use of "Immobilines" [117]. Thus, the monitoring of the pH of a zone as a function of electrofocusing time is central to gel electrofocusing. Using gel tube apparatus it is easily carried out: At the desired time periods, a gel tube is withdrawn from the apparatus and analyzed for protein and pH. Using gel slab apparatus of the commercial type, such an analysis is far more problematic. Here an analysis of protein patterns as a function of electrofocusing time can only be carried out either by slicing off of channels from the gel slab at the desired time intervals, or by applying the sample to the various channels in a staggered fashion. The first is hardly possible without disrupting the adherence of the gel to the glass walls. The second approach suffers from pH gradient instability

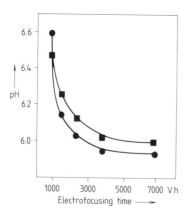

Figure 31. Representative dynamics of pH positions of two proteins as a function of the time of EF (in Volt hours). Note the asymptotic approach towards a constant pI position characteristic for each protein. Data of [188].

with time – samples applied at different electrofocusing times are in fact separated in non-identical pH gradients. This problem may vanish with stabilization of pH gradients by Immobilines [117] or otherwise [191],[192]. At this time, neither approach has been particularly successful when applied across a wide pH range. Vertical gel apparatus also does not allow for a pattern analysis as a function of electrofocusing time, for the same reasons as stated above for common designs of horizontal slab apparatus.

The apparatus suggested for use in gel electrofocusing is identical to that used for PAGE (Chapter 4) with the sole exception of an electrode arrangement which allows one to regulate the voltage gradient across the gel only (Fig.32). One electrode is introduced through a hole in the electrode holder top above the gel tube and is allowed to touch the top of the gel; the other is seated centrally in the apparatus, and can be adjusted in vertical position so as to touch the bottom of the gel tubes. There are both empirical [118],[119[and theoretical [20],[115] reasons to use the leading and trailing carrier constituent as anolyte and catholyte, in order to enhance the stability of natural pH gradients (see Section 12.4). This brings about high voltages across the electrolyte reservoirs which are variable with electrofocusing time. This variability makes it impossible to control the voltage across the electrolyte reservoirs and the gel during electrophoresis. To achieve a control of voltage across the gel, electrodes must be brought into contact with the top and bottom of each gel (positioned in the upper electrolyte reservoir). This has been achieved by a circular bottom electrode the height of which can be adjusted to the level of the gel tube bottoms, and by an upper electrode manifold that allows one to lower individual electrodes into each of the tubes, to the level of the gel [120].

Of particular importance to gel electrofocusing with weakly acidic and basic anolytes and catholytes is the large size of the reservoirs of the apparatus. This is 500 and

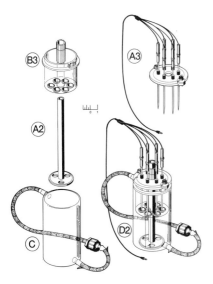

Figure 32. Gel tube apparatus for EF [120]. Dimensions are given in inches.

3000 mL in the case of tube apparatus I, and at least a few hundred mL in models II-IV (Section 4.5). These volumes suffice to provide the necessary buffering capacity. Small volume reservoirs, and even more so filter paper electrolyte strips in lieu of reservoirs, make it impossible to take advantage of weakly acidic and basic anolytes and catholytes for the purpose of tailoring the pH gradient to one's needs, using "constituent displacement" [121], and for the purpose of pH gradient stabilization [118]. Thus, many models of horizontal slab apparatus presently available are restricted to the application of strongly acidic and basic anolytes and catholytes, with the corresponding limitations with regard to gradient stability in EF.

Certain apparatus conditions in electrofocusing relate to the lability of EF gels which is due to the requirement for "non-restrictiveness" (in common with stacking gels) and the fact that gels hydrate in pH dependent fashion, giving rise to variously swollen diameters along the pH gradient. The common failure of vertical gel slab EF is probably due to unevenness of wall adherence and therefore field strength across the various channels of such slabs (frequently made of soft glass rather than Pyrex), in the absence of measures securing firm and uniform wall adherence.

The remedies for mechanical gel instability are the same as in MBE on gels: Coating of the tube walls with 1% Gelamide 250 or agarose, mechanical support of the gels with Nylon mesh and care not to stress the gel mechanically or by hydrostatic pressure (see Section 4.5.3 for suitable procedures). Moreover, there is a test for wall adherence that is not available for MBE: The gel tube is once suspended in the anolyte, once in the catholyte. pH gradients as a function of focusing time are measured. They will be superimposable in the 2 cases if wall adherence is adequate; if it is not, positioning the gel into the (acid) anolyte will shift the pH gradient toward acid; positioning it in the (basic) catholyte will shift it in the basic direction [37]. This important wall adherence test in EF necessitates vertical apparatus.

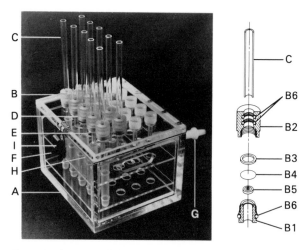

Figure 33. Apparatus for packing multiple 6 or 18 mm ID gel tubes with a granulated gel such as Sephadex G-75 [54].

In addition to "non-restrictive" polyacrylamide, agarose [122], granular dextran (Sephadex) [123],[54],[99] and agarose-polyacrylamide copolymer (LKB, Ultrogel) [100] matrices are compatible with EF, and particularly useful preparatively. These gel media are most often applied in flat-bed apparatus. The usual limitations of such apparatus with regard to temperature and moisture control has been overcome by some recent designs described above (Section 4.4). The limitation of such apparatus with regard to sample volume does not apply to EF, since the sample can be uniformly distributed throughout the gel bed and be used in its hydration.

Granular media are also compatible with the tube apparatus (Section 4.2) [54] and can therefore be applied in those cases where stringent temperature control is required [99],[120]. To pack a granular medium into gel tubes, it is convenient to apply a vacuum. An evacuated packing chamber for multiple tubes of 6 and 18 mm ID has been described [54],[56] and is depicted in Fig.33. After packing, a polyacrylamide plug is affixed to the ends of the gel. The sample is dispersed into the granular medium prior to formation of the upper plug. Otherwise, only the procedure of gel removal from the tube and of gel analysis varies from that which applies to polyacrylamide (see below).

12.2 Gel

The approach of proteins toward their isoelectric positions is slow compared to the carrier ampholytes due to their lower charge density and mobility. If in addition, the migration of the protein toward its isoelectric position on the pH gradient is slowed through molecular sieving, the likelihood of reaching that position, and isoelectric separation from other proteins, within the finite lifetime of the pH gradient diminishes. It is therefore imperative in electrofocusing to use a "non-restrictive" gel. The conceptual limitations of that term, and the mode of determining it experimentally, are the same as for MBE (see Section 7.8). As in preparative MBE, an attempt is made to combine "non-restrictiveness" with the greatest possible degree of wall adherence. Furthermore, due to pH dependent gel swelling in electrofocusing, an elastic gel is desirable. Both criteria favor the use of 5 %T, 15 %C_{DATD} gels for electrofocusing at 0° C of species with molecular weights less than a few hundred thousand.

The choice of a gel matrix in electrofocusing also involves, in some applications, a consideration of protein aggregation which in this method is promoted by its low operative ionic strength and by the isoelectric state of the protein. This protein aggregation is concentration dependent, proceeding in the representative case of glucocorticoid receptor progressively, as the polyacrylamide gel concentration is raised [52]. Aggregation is therefore promoted by gel matrices which give rise to either high protein concentration at a restrictive gel surface, or which are relatively effective in lowering zone diffusion, giving rise to relatively sharp (concentrated) zones. The latter condition holds particularly when electrofocusing is prolonged beyond the minimal time needed to approximate the isoelectric endpoint: Due to aggregation and precipitation, protein recovery decreases progressively with time of electrofocusing under those conditions. Examples for the need to consider aggregation in the choice of gel matrix are the preference for Sephadex G-75 (Pharmacia) over polyacrylamide, agarose or Ultrogel AcA 34 (LKB) in the case of the glucocorticoid receptor [99],[111],[52], or the preference for Ultrogel AcA 44 (LKB) over Sephadex G-75 in the case of protein kinase [100].

Agarose (Iso-Gel) is used commonly in flat-bed apparatus [122], but appears frequently to be an insufficiently stable matrix for electrofocusing in gel tubes. It seems unlikely that wall adherence of agarose to agarose coated Pyrex tubes would be substantially weaker than that of polyacrylamide, or that pH dependent local gel swelling and shrinking would be greatly different. It is more likely that the apparent advantage of flatbed apparatus for agarose EF may relate to the relatively low potential gradients (V/cm) employed in the flatbed as compared to the gel tube technique, and that therefore local Joule heating due to uneven voltage across the pH [124],[125] is less prone to melt the gel.

12.3   Selection of a Carrier Constituent Mixture

12.3.1   Theoretical Working Hypothesis

In the selection of carrier ampholytes, as later in the selection of anolyte and catholyte, rationales will be presented which are based on the hypothesis that EF is a special case of MBE, characterized by multiple sequential moving boundaries and by the slowing down ("arrest") of these boundaries when the solvent ions (proton and hydroxyl ion) become the sole common ions of the system [20],[115]. Support for that hypothesis derives from the approximate fit of experimental pH gradients to the ones predicted by the MBE theory [115] and, indirectly, from agreement of predicted with empirically observed conditions for pH gradient stabilization [116],[195]. It should nonetheless be kept in mind that the fit of data to a hypothesis can never prove more than the compatibility of the model with the known facts – it cannot prove that the model represents reality. It should also be realized that the isoelectric pH, at which proteins are found at the steady-state in EF, in no way contradicts the notion that pH gradients form by an MBE mechanism. When solvent ions become the sole counterions of the protein, the proteins which are the trailing ions in their respective moving boundaries are arrested through operation of the moving boundary equation; proteins with zero net mobility necessarily are at an isoionic pH. The fact that natural pH gradients form from non-amphoteric constituent mixtures [115],[116],[126] shows certainly that pH gradients can form by a mechanism other than isoelectric condensation of constituents. The fit of experimental pH gradients to theoretical prediction by an alternative theory of Bier et al. [127] compatible with an isoelectric condensation mechanism for carrier constituents cannot be verified at present because its mathematical complexity rules out on economic grounds the computations required for systems of more than a few carrier constituents. (That theoretical treatment of pH gradient genesis is not to be confused with the one by Svensson concerned with the isoelectric condensation of ampholytes in a pre-existing pH gradient [128]).

To predict a natural pH gradient, i.e. a pH gradient formed by the electric field alone through an alignment of carrier constituents in the field in order of their net mobilities (or, in case of ampholytes, pI), one needs a mixture of defined constituents varying in pK and ionic mobility [115],[195]. Since for most carrier constituents ionic mobilities are not known at this time, carrier constituents must frequently be selected empirically in "buffer EF" (Section 12.3.2) on the basis of their known pK and pI values (Fig.34) or, in the case of synthetic carrier ampholyte mixtures (SCAMs) (Section 12.3.3), their pI-ranges (Fig.35A). In contrast to natural pH gradients are those which are produced mechanically by use of gradient mixers (Section 12.3.4). In that category, gradients can be predicted on the basis of the pK values and molarities of Immobiline monomers [117],[129], i.e. vinyl compounds substituted to various degrees with amino and carboxyl groups. In addition to the pKs, pIs or pI ranges of carrier constituents, their molecular size (Section 12.3.5) and their conductance properties (Section 12.3.6) need to be considered. The empirical choice of carrier constituents is further governed by the following considerations. The pH range produced by the particular mixture of carrier constituents should be such that the isoelectric pH of the protein be located at the center of the pH gradient. At least in principle, the pH gradient should be flat around the isoelectric point. To be smooth and linear, a mixture of the greatest number of carrier constituents should be selected.

12.3.2 Buffer EF

In principle, the selection of carrier constituents which are simple buffers of known chemical identity can produce pH gradients tailored to the particular requirements of a separation. In practice, however, it is not possible to predict pH gradients on hand of a list of pKs and pIs of suitable buffers (Fig.34) in better than qualitative fashion. The reason that pH gradients cannot be predicted from pIs and pKs of the buffers is that the alignment of carrier constituents in the pH gradient and their phase compositions are a function of their net mobilities which are derived from not only pKs but also from the ionic mobilities of the buffers [115],[195].

Empirically selected mixtures of acidic carrier constituents differing in pK or pI yield acidic pH gradients, those with basic ones basic pH gradients and those with neutral constituents neutral pH gradients [115],[126],[130]. Obviously also, the addi-

Buffer Electrofocusing Carrier Constituents

Amphoteric Constituents	Abbreviation	pI[a]	pK_2[a]	pK_1[a]	Concentration (mM)	Supplier[b]
1 L-Glutamic acid		3.0			10	S
2 2-Pyridine carboxylic acid (Picolinic acid)		3.2			10	S
3 4-Pyridine carboxylic acid (Isonicotinic acid)		3.3			10	S
4 3-Pyridine carboxylic acid (nicotinic acid)		3.4			10	S
5 2-Aminoethane sulfonic acid (taurine)		5.1			20	S
6 D,L-Threonine		5.6			10	S
7 Glycylglycine		5.7			10	S
8 Glycine		6.0			20	S
9 N-Methylglycine (sarcosine)		6.1			20	S
10 L-Proline		6.3			10	S
11 Iminodiacetic acid		6.4			20	S
12 Creatine		6.8			20	S
13 γ-Amino-n-butyric acid	GABA	7.3			20	S
14 ε-Amino-n-caproic acid	EACA	7.6			10	S
15 L-Histidine		7.7			10	S
16 Guanidoacetic acid (glycoyamine)		8.0 est.			20	S
17 N$^{\alpha}$-β-Alanyl-histidine (carnosine)		8.2			20	S
18 L-Lysine monohydrochloride		9.7			20	S
19 L-Ornithine monohydrochloride		9.7			20	S
20 L-Arginine		10.7			10	S
21 L-Asparagine		10.7			20	S
22 Glycine betaine (lycin; betaine)				1.8	20	S
23 γ-Amino-β-hydroxybutyric acid betaine (D, L-carnitine hydrochloride)				3.9 est.	20	S
24 2-(N-Morpholino)ethane sulfonic acid	MES		6.1		20	S
25 N-(2-Acetamido)-iminodiacetic acid	ADA		6.6		10	S

Figure 34. Buffers suitable for EF [35].

[a] At 25 °C[c]
[b] A: Aldrich Chemical Co., Milwaukee, WI
B: Baker Chemical Co., Phillipsburg, NJ
F: Fluka/Tridom Chemical Inc., Hauppauge, NY
R: Research Organics, Inc., Cleveland, OH
S: Sigma Chemical Co., St. Louis, MO
[c] Synthesized from biguanide sulfate

Amphoteric Constituents	Abbreviation	pI[a]	pK_2[a]	pK_1[a]	Concentration (mM)	Supplier[b]
26 N-(2-Acetamido)-2-aminoethane sulfonic acid	ACES	6.8			20	S
27 3-(N-Morpholino)-2-hydroxypropane sulfonic acid	MOPSO	6.9			10	R
28 N, N-Bis(2-hydroxyethyl)-2-aminoethane sulfonic acid	BES	7.1			10	S
29 3-(N-Morpholino)-propane sulfonic acid	MOPS	7.2			10	R
30 N-Tris(hydroxymethyl) methyl-2-aminoethane sulfonic acid	TES	7.4			20	S
31 N-2-Hydroxyethylpiperazine-N'-2-ethane sulfonic acid	HEPES	7.5			20	S
32 3-[N-bis(Hydroxyethyl)-amino]-2-hydroxypropane sulfonic acid	DIPSO	7.5			10	R
33 3-[N-(Tris-hydroxymethyl-amino]-2-hydroxypropane sulfonic acid	TAPSO	7.6			20	R
34 N-Hydroxyethylpiperazine-N'-2-hydroxypropane sulfonic acid	HEPPSO	7.8			10	R
35 N-2-Hydroxyethylpiperazine propane sulfonic acid	EPPS	8.0			10	S
36 N-Tris(hydroxymethyl) methyl glycine	TRICINE	8.1			20	S
37 N,N-bis(2-hydroxyethyl) glycine	BICINE	8.3			20	S
38 N-Tris(hydroxymethyl) methyl)methyl-3-amino-propane sulfonic acid	TAPS	8.4			10	S
39 3-[N-α, α-Dimethyl hydroxy-ethyl)-Amino]-2-hydroxy-propane sulfonic acid	AMPSO	9.1			10	S
40 2-(N-Cyclohexylamino) ethane sulfonic acid	CHES	9.5			10	S
41 Cyclohexylaminopropane sulfonic acid	CAPS	10.4			10	S
Non-amphoteric constituents						
42 Malic acid				3.4	10	S
43 Lactic acid				3.9	10	S
44 Acetic acid				4.8	10	B
45 Propionic acid				4.9	10	S
46 Benzamidine				11.6	20	F
47 Biguanide			13.2	3.1	10	A[c]

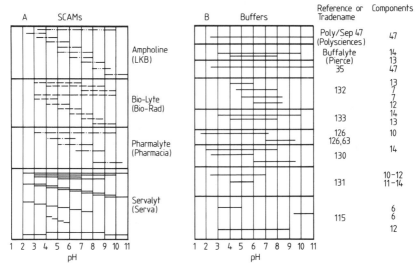

Figure 35. pH Ranges of (A) commercially available synthetic carrier ampholyte mixtures (SCAMs) and (B) buffer mixtures in EF.

tion or subtraction of constituents in each of these 3 broad categories has the expected effect in making pH gradients more acidic, more basic or more neutral [130],[131]. However, such additions may lead to salt forms between acidic and basic carrier ampholytes. Addition of a basic constituent may modify the acidic end of the pH gradient [35]. The addition of the components of a basic pH gradient to those of an acidic one does not produce the sum of the two pH gradients at the basic end [115]. All pH gradients spanning neutrality must give rise to such charge interactions between stacks of anionic and cationic constituents while they cross one another during their boundary displacement in opposite directions. The charge interaction products persist in the electric field for time periods practically required for the attainment of the steady-state positions by proteins. It is therefore misleading to think of such interactions as being transient in a practical sense – although over long time periods they may be. If one wishes by mere choice of carrier constituents to construct exactly the desired pH range, and if one wants to stabilize it for a sufficient time period, tedious systematic experimentation is needed at this time. It is therefore advisable in practice to utilize those buffer pH gradients which have been reported in the literature or which are commercially available (Fig.35B). Thereby, necessarily, buffer pH gradients, like their SCAM counterparts, cease to be "tailor made", unless one applies "constituent displacement" (see Section 12.4) to them. Nonetheless, simple buffer constituents have advantages over SCAMs, which may make them preferable: a) Known chemical properties; b) dialyzability; c) minimal interactions with proteins (see Section 12.3.4); d) inexpensiveness of mixtures if they are prepared in the laboratory.

12.3.3   EF in SCAMs

With commercial SCAMs, the number of obtainable pH ranges is practically limited to the available preparations and their mixtures (Fig.35A). "Home-made" SCAMs [134],[135] could be prepared to measure by empirical, systematic variation of the amine and acid composition of the polymerization mixture, as used in the preparation of limited SCAM pI ranges by Pharmacia (Pharmalytes). Any desired pI range can be prepared by electrofocusing a given mixture and isolating the pH region of interest. This procedure is used commercially, but is prohibitively expensive in the laboratory.

When a desired pH gradient is prepared by mixing the various commercial pI ranges, attention must be paid to carrier constituent interactions to form stable salts, as pointed out in Section 12.3.2. Particularly, the high pI range SCAMs frequently added to the protein sample to ensure its solubility not only extend the pH gradient into the sample phase but also are likely to perturb the acidic end of the gradient by electrostatic binding of acidic SCAMs.

An alternative mode of constituent selection (designated as "constituent displacement") consists of withdrawing from the aligned constituents all components with pIs less than the anolyte or more than the catholyte, by controlling the pH of (large volumes of) the anolyte and catholyte (see Section 12.4). Similarly, the slope of pH gradients can be modified by choice of the anolyte at pH values outside the pH range of the gradient (see [119] and Section 12.4).

In general, the selection of carrier constituents is linked with the problem of pH gradient instability. A particular mixture will not give rise to a single pH gradient, but to different pH gradients as a function of electrofocusing time. Thus, the selection process already requires an evaluation of pH gradient at various electrofocusing times.

12.3.4   Immobiline

The selection of pK values and mixing ratios in the use of Immobiline is done by aid of either a nomogram (LKB Application Note 321) or by use of computer programs [129]. Narrow pH gradients across one to three pH units are much more readily available by this method than a pH gradient covering the entire useful pH range. This may in part be due to the greater instability of basic vinyl derivatives due to catalysis of polymerization by the basic groups. The central promise of Immobiline pH gradients is their temporal stability [117], excelling even over that found with flat pH gradients in general. Their second important feature is the possibility, from a knowledge of pKs alone, to predict and tailor any pH gradient to one's needs. Unlike in the selection of the other types of carrier constituents, however, attention must be paid to the rates and extent of incorporation of the various vinyl monomers used into the gel.

12.3.5   Molecular Size of Carrier Constituents

The molecular sizes of all SCAMs used to set up natural pH gradients are larger than the sizes of simple buffers or amino acids (MW 150–200). A recent estimate is MW 600–900 for Ampholine, 400–700 for Servalyte and 300–600 for Pharmalyte [136]. Due to their relatively large molecular weights, SCAMs exhibit hydrophobic and H-bond binding to proteins and to one another to varying degrees, while simple buffers used to set up natural pH gradients do not [137]. The binding of basic SCAMs to acidic species such as heparin [138] or nucleic acids [135] is best documented; it seems equally likely for basic proteins [38]. This type of binding is urea-dissociable and therefore at least in part due to H-bonding. SCAM interactions with one another or with proteins also involve hydrophobic bonds since they are sensitive to such solvents as dioxane or DMSO [139]. Into that category fall the binding of SCAMs to Triton X-100 [139] and to hydrophobic membrane proteins such as cytochrome P-450 (L. Hjelmeland, personal communication).

In Immobiline pH gradients, no binding studies have been reported to date. It is to be expected, however, that binding of proteins to the carrier ampholytes should be greater even than with SCAMs in view of the larger size of the polymeric carrier ampholyte species.

12.3.6   Conductance Properties of Carrier Constituents at the Steady-State

Constituents should be selected to provide an even conductance across the pH gradient, to ensure that carrier constituents and proteins are not retarded during passage of low voltage regions and thus are able to migrate to their steady-state positions in the time allotted, and to protect the proteins from encountering local "hot spots" during their passage through high voltage regions. Since conductance of acidic and basic carrier ampholytes at their pIs is necessarily higher than that of the neutral species [136], a balanced conductance across the pH gradient can be achieved either by increasing the concentration of neutral carrier constituents relative to the acidic and basic species, or by selection of neutral carrier ampholytes with small pI-pK differences which conduct relatively well at their pIs [128]. In this regard, a small pI-pK difference is 1–1.5, a large difference is 4-5. Using SCAMs, which have been designed to possess small pI-pK values and therefore good conductance at the pI [140], this can only be done by enriching the system with neutral pI ranges.

Buffer EF is restricted to date to carrier constituents with relatively large pI-pK values (Fig.34). The pH gradients in buffer EF on polyacrylamide gel containing 10–20 mM carrier constituents are therefore less conductive at the steady-state, by one order of magnitude, than those formed by 1 or 2% SCAMs [35]. [In Sephadex EF steady-state currents similar to 1% SCAM have been observed with Poly/Sep 47

(Polysciences). However, the currents during the formation of the pH gradient from buffer mixtures are much greater with Poly/Sep 47 compared to SCAMs at the same voltages, and the transient state of pH gradient formation persists for much longer times.] Potentially, at least, buffer pH gradients can be adjusted to an even conductance across the gradient by adding neutral constituents at augmented concentrations to the constituent mixture.

Immobiline electrofocusing allows one to compute equiconducting pH gradients [129]. However, these remain to be demonstrated over a wide pH range. Experimentally, such wide pH gradients prepared according to the published recipes appear to yield variable conductance across the pH gradient, with negligible voltage and mobility at the extremes of pH, consequently prolonged electrofocusing times and a high degree of lateral zone diffusion. Another problem associated with high resistance across the neutral pH region is the difficulty in measuring pH by the glass (contact) electrode.

To date, EF has been used nearly exclusively under conditions of an extremely low ionic strength at the steady state. This has greatly increased the likelihood of ionic interactions between proteins, carrier constituents or those two among each other. It must also lead to protein solubility problems, since solubility decreases and the pH range for isoelectric precipitation increases with a lowering of the ionic strength [1]. In the EF of hydrophobic proteins, which are frequently bound to other membrane components via their hydrophilic domains, low ionic strength may produce large molecular weight complexes [1]. From all of these viewpoints, and considering the predilection of many proteins for physiological ionic strength, EF at high ionic strength would be desirable. It is, of course, possible to increase carrier constituent concentrations, to add salt to all phases [193] or to use buffer salts as the catholyte and anolyte, thereby increasing the flow of current through the stack [146] to increase the operative ionic strength of the system. However, with present apparatus, this leads to wattages which cannot be dissipated sufficiently to prevent heating if a convenient experimental time scale (not exceeding 1–2 days) should be maintained. Thus, the problem of high ionic strength EF is one of improving the heat dissipation capacity of the apparatus through Peltier cooling, thin apparatus walls and the choice of heat-conductive apparatus materials (Chapter 4).

12.4 Selection of Anolyte and Catholyte

Commonly, no attention is paid to the choice of anolyte and catholyte in electrofocusing other than to select a strong acid as the anolyte, a strong base as the catholyte, usually at equivalent concentrations. The selection of the appropriate anolyte and catholyte is, however, important for a number of reasons: a) the stabilization of pH gradients as a function of anolyte/catholyte pH [119] and anolyte/catholyte concentration [118]; b) the tailoring of the pH range to one's needs by "constituent displacement"; c) the reversal of the pH gradient drift as a function of the relative pHs of anolyte and catholyte [37].

12.4.1   Stabilization of pH Gradients by Selection of Anolyte and Catholyte

In theory [20],[115], the catholyte and anolyte in EF should have the composition and pH of the leading and trailing phases of the system of sequential moving boundaries which gives rise to the pH gradient. If that condition is met, the steady-state alignment is not perturbed by the presence of large amounts of the leading and trailing ions in their respective terminal positions. If catholyte and anolyte contain ionic species with net mobilities less than that of the leading ion or more than that of the trailing ion, the steady state cannot be reached in finite time, i.e. the pH gradient will remain transient. This consideration is, of course limited to a single, either cationic or anionic, train of carrier constituents, and to pH gradients which extend to neutrality, but do not encompass both the acidic and basic pH range. When the pH gradient encompasses the entire pH range, both a cationic and an anionic train of constituents align at the steady state. Now the leading phase of the cationic train is strongly basic, i.e. the hydroxyl ion concentration in the catholyte is high. Correspondingly, the proton concentration in the anolyte is high. Considering that the mobilities of proton and hydroxyl ions are higher than those of any of the other constituents, the anolyte becomes a source of protons passing continuously through the stack to align itself as the leading ion of the cationic train, while the hydroxyl ion must do the same to occupy its position as the leading ion of the anionic train. Thus the steady-state is continuously perturbed and the train of constituents is more highly ionized, leading to an increased boundary displacement rate. Consequently, one would predict a pH gradient instability in proportion to the acidity of the anolyte and the basicity of the catholyte. A third theoretical prediction is that the displacement rate of a moving boundary decreases with increasing concentration of the leading ion, and that therefore pH gradient stability can be increased by selection of a catholyte and anolyte at high concentration.

Experimental evidence confirms these 3 predictions. a) pH gradients are stabilized when the catholyte and anolyte compositions are identical to those of the leading and trailing phases. In buffer EF, this is achieved by using the leading and trailing constituents as the anolyte and catholyte [191]. Note that in an acidic pH gradient, both anolyte and catholyte may be acids [130]. In SCAM EF, this can be achieved either by using the SCAM as anolyte and catholyte, or by addition of a leading and trailing buffer constituent to the SCAM, and using these buffers as anolyte and catholyte [118],[192]. b) The lower the acidity of the anolyte, the more stable is an acid-to-neutral pH gradient [119]. Neutral-range pH gradients with only mildly acidic and basic anolyte and catholyte are the most stable [118],[130]. Stability of pH gradients is enhanced when catholytes and anolytes are isoelectric ampholytes, or acids and bases 2 pH units or more below and above their pKs and thus effectively uncharged. Since such solutions have little buffering capacity, relatively large buffer reservoirs are required and monitoring of pH and conductance of these reservoirs during EF becomes necessary. If their buffering capacity is significantly exceeded and pH should change beyond the desired pHs of the gradient termini, the anolyte and catholyte need to be replaced intermittently during EF. The corollary of this fact with regard to apparatus construction has been discussed in Section 12.1. Obviously, when buffer reservoirs are small and anolytes/catholytes of limited solubility are used, electrolysis will give rise to pH changes such that independently of the buffer species used, strongly acidic and basic anolytes/catholytes are produced, and no effect of the choice of anolyte/catholyte on the pH range of the gradient is exerted [141]. c) High concentrations of anolyte and catholyte have a stabilizing effect [118]. Usually, at least 0.1 M anolytes and catholytes are used.

12.4.2 Constituent Displacement

The choice of anolyte at a pH within a given pH gradient has been used to displace carrier ampholytes and proteins positioned between the terminus of the gradient and the selected pH ("constituent displacement") [121],[144]. As pointed out above, this technique depends on constant anolyte and catholyte pHs and therefore (since these are amphoteric or 2 pH units removed from the pK, i.e. poorly or not buffering) on large electrolyte volumes and reservoirs. It should also be noted that the system is in a transient state while displacement proceeds, and that therefore the zone sharpness of regulated phases cannot be maintained during displacement, with consequent partial loss of resolution.

12.4.3 Drift Reversal

It has been generally assumed that the "drift" of pH gradients during pH gradient decay [142],[143] is always in the cathodic direction. However, this is not the case. Experimentally it has been found that when the pH of the catholyte relative to that of the anolyte is sufficiently increased, a pH gradient drift in the opposite direction occurs [37]. This finding is in agreement with theory [20] which regards "pH gradient instability" as being synonymous with boundary displacement. Since the relative mobilities of cations are larger than those of anions, the train (stack) of constituents displaced towards the cathode is usually more rapidly displaced than the opposite train in the direction toward the anode. However, this situation can be reversed in a moving boundary system with strongly basic and weakly acidic leading phases. A strongly basic catholyte in that case must increase the displacement of the anionic train of constituents by making OH⁻ the leading ion, and by decreasing the ionisation of the cationic and increasing that of the anionic stack. It should also be noted that the drift reversal by selection of a basic catholyte and weakly acidic anolyte is incompatible with the notion that the drift mechanism is electroendosmotic [145]. (The demonstration that electroendosmosis purposefully produced by incorporation of net charge into a polyacrylamide matrix can set up a "drift" does in no way prove that the drift in electroendosmosis-free polyacrylamide gels proceeds by that mechanism.)

12.5   Procedure of Preparing the Electrofocusing Gel

12.5.1   Polyacrylamide

Electrofocusing gels made of polyacrylamide are prepared by the procedure given in Section 5.5). The gel buffer is replaced by a 4-fold concentrate over the final concentration of the carrier ampholyte mixture, e.g. a 4% SCAM preparation, if EF is to be conducted in a 1% SCAM, or a 0.08 M buffer mixture, if it is to be conducted in 0.02 M buffer carrier constituents. The sole specific procedural requirement of the formation of polyacrylamide EF gels refers to the fact that many basic carrier constituents act as polymerization catalysts, presumably in a manner akin to TEMED. Thus, if the electrofocusing gel polymerizes faster than in 10 minutes, catalyst concentrations must be reduced or inhibitors such as K-ferricyanide must be introduced to decrease the polymerization rate.

EF requires a "non-restrictive" gel, and therefore the choice of gel concentration, crosslinking agent and degree of crosslinking is governed by the same considerations as those pointed out for stacking gels (Section 7.8.2). However, the need for mechanical gel stability and good wall adherence is exacerbated in EF compared to stacking gels, since the variable pH along the gel length brings about variable gel hydration (swelling) along the gel. Necessarily, this problem is more severe in wide than in narrow pH gradients. In general, 5 %T, 15 %C_{DATD} gels appear to be the best compromise solution to the conflicting demands of "non-restrictiveness" and mechanical stability [80] in EF at 0 °C.

The stability of pH gradients increases with the concentration of carrier constituents [126],[143], as does the time required to establish the steady-state with regard to the carrier constituents, i.e. the steady-state pH gradient. A reasonable compromise value appears to be 0.1 M buffers or 2% SCAMs in many cases.

12.5.2   Granular Gel

If a granular gel matrix such as Sephadex G-75 is used as a matrix for electrofocusing in gel tubes, the suspension containing carrier constituents and, optionally, the protein sample, is poured into tubes on the evacuation device (Fig.33) and subjected to a house vacuum (about 50 micron). The filtrate is collected and reapplied if it contained sample. The bottom few mm of bed are removed from the tube by spatula and filled with a sturdy support gel made in anolyte (e.g. a 10 or 20 %T, 2 %C_{Bis} gel). Then the top of the gel tube is filled with the same gel made in catholyte [54].

12.6   Preparation of the Sample

12.6.1   Weakly Acidic/Basic Anolyte/Catholyte

When the cathodic end of the gel is loaded, and the catholyte is identical to the terminal cathodic constituent of the pH gradient, the sample is made in 0.01 to 0.03 M catholyte, in 10–20% sucrose, and containing some dye for purposes of visualization during sample application. However, when the catholyte pH is close to the pI of the protein, this procedure may result in aggregation and isoelectric precipitation of the sample before electrofocusing commences. In this case, it is necessary to raise the pH of the sample as much as possible within the constraints of denaturation. E.g. in the case of human growth hormone, the preparation of the sample in 0.1 M arginine was found to be appropriate [130]. Similarly, if a SCAM pI range 3.5–10 with pH 5.5 to 6 is used as a carrier constituent preparation, many proteins would aggregate and precipitate during the few minutes before the current is turned on, unless one prepares the sample in 1% pI range 8–10 or 9–11 SCAMs. Such sample phases will of course make the cathodic gel terminus more alkaline at the steady-state and cause pH gradient shifts due to the formation of salts with strongly acidic SCAM species, but their overall effects on the final pH gradient are usually negligible if the alkaline SCAMs are introduced in a small amount relative to the other pI ranges used. However, at least in the case of an anion-binding protein, basic SCAMs have been found to complex, thus shifting the pI′ upward; this binding is aggravated when the SCAM preparation is not fresh (and therefore presumably aggregated) [195].

12.6.2   Strong Acid/Base Anolyte/Catholyte

If strongly acidic and basic anolytes and catholytes are used which will tend to denature the sample prior to electrofocusing, it is necessary to prepare the sample in a well-buffering solution of carrier constituents at 30% sucrose concentration, and to overlayer the sample by a protective layer of the same carrier constituent solution in the absence of the protein and at 10% sucrose concentration. This protective layer is then overlayered with the adjacent electrolyte with care not to disturb the interface. Thus, in this case the sample is pipetted directly onto the gel, rather than underlayered.

12.6.3   Volume

Sample volumes in gel electrofocusing may be rather large, although not as large as in MBE where current density is sufficiently large to allow for relatively rapid attainment of the steady-state. A 1.7 mL EF gel in a 6 mm tube may tolerate 100–300 μL of sample. Gel electrofocusing on granular gel beds excels in load capacity over that in all other media since it allows for the hydration of the gel matrix material by the sample.

12.7   Procedure of Electrofocusing

The procedure of gel electrofocusing in gel tubes is identical to that detailed in Section 5.5 (polymerization) and Section 7.6 (electrophoresis), with the following exceptions:
- Voltage regulation
- Determination of the endpoint of electrofocusing.

12.7.1   Voltage Regulation

During the alignment of carrier constituents in order of their net mobilities (the first transient state of the pH gradient with regard to the carrier constituents) the current is very high initially, and decreases asymptotically. To prevent undue Joule heating of the sample during this phase, the current should be regulated at the same levels used in PAGE, i.e. 1 mA/tube at 0 °C. Once the current has reached its asymptotic value, current control is replaced by voltage regulation. When strong acid and base are used as anolyte and catholyte (and therefore the voltage drop across the reservoirs is negligible), this change to voltage regulation is carried out once the voltage has risen to 200 V (20–40 V/cm of gel). Usually, this is the case within 1 h of electrofocusing. If weakly acidic and basic anolyte and catholyte are used, the voltage across the gel must be controlled at the same levels using an electrode arrangement which allows one to position electrodes at the top and bottom surfaces of each gel (see Section 12.7.2).

As an alternative to the change from regulated current to regulated voltage, it is in principle possible and preferable to regulate the wattage. However, since 0.1 to 0.25 W can be dissipated per gel tube of 6 mm diameter, and present wattage controlled power supplies are incapable of regulating fractions of a watt, this is only practical today if many gel tubes are run simultaneously. Even if one starts to electrofocus with many tubes, one usually terminates the experiment with a single tube (see Section 12.7.3). Thus, in the middle of the run, one would have to relinquish wattage control for regulated voltage. Thus, one does not gain over the procedure proposed above; rather it is much easier to switch to voltage control 1 h after the beginning of the run than at some indeterminate end stage of the experiment.

12.7.2   Determination of the Endpoint of Electrofocusing

Commonly, the endpoint of an electrofocusing experiment is taken at the time when an asymptotic current level is reached. That time indeed ends the first transient state of EF, but only in regard to the carrier constituents, not in regard to the proteins.

One way to recognize the attainment of the steady-state by the protein is to apply it throughout the gel (if a granular gel is used) or at both the anodic and cathodic termini (or at any other pHs on both sides of the pI). Then, the end point is defined as the time at which the cathodically and anodically migrating zones of the protein fuse. Procedurally, this is easily done on horizontal gel slabs, if the protein of interest can be detected at various times in the same channel by use of the filter paper blotting technique [33]. It is far more difficult by excision and staining of slab channels at various times, since the excision easily disrupts the wall adherence of the entire gel slab. With gel tubes, the technique of bilateral loading requires application of one of the samples in a small dialysis bag attached to the bottom of the gel; the bag is held to the tube by a sleeve of Tygon tubing. This operation, however, is too artful to be generally useful.

The easiest way to determine the attainment of the steady-state pI position for the protein of interest in gel tube electrofocusing is to withdraw tubes from the upper buffer reservoir periodically, e.g. every 3 hours, to measure the pH gradient on each gel, to stain or assay the same gel, and to plot the pH at which the protein is located as a function of time. A constant pH value – the apparent pI (pI′) – is being approached asymptotically with time (Fig.31). The asymptotic pH value is by definition the pI′, and the correct focusing time is the one required to reach this pI′.

If no protein activity is available by which the protein of interest can be detected, the same experiment can be carried out to determine the time at which pattern constancy is attained [147]. After that time, the "Gestalt" of the stained gel pattern remains the same, but the pattern is being progressively compressed by the usually cathodic drift of the pH gradient. Ultimately, the proteins migrate in order of their relative basicity into the catholyte [147].

Whether the isoelectric end point for the protein of interest is determined by bi-directional migration and zone coalescence, or by a curve of pH *vs* electrofocusing time, or by recognition of the time at which pattern constancy is reached, it is essential to realize that the isoelectric end point varies from protein to protein. Thus, it is wrong to determine it by some arbitrary colored protein for the other proteins in the system as is being commonly done.

12.8   Gel Evaluation

12.8.1   pH Gradient Analysis

The pH gradient is a *dynamic* entity: The first transient state [148], in which first the carrier constituents, then the proteins reach the steady-state, is followed immediately by the second transient or decay state during which (if the drift is cathodic) the stack of carrier ampholytes and proteins is displaced from the gel into the catholyte (Fig.36). Thus, there is no steady-state with regard to the positions of the carrier ampholytes and proteins on the gel; there is only a steady-state with regard to pH: Proteins do approach a constant pH; so do buffer carrier constituents [115]. With SCAMs, we cannot assay the individual carrier ampholyte species. Work with pH gradients formed by labelled amino acids has shown that the pH positions of constituents need to be tested in an electrofocusing gel with even conductance across the gel. In the absence of even conductance, amino acids tend to accumulate as peaks at those points on the gel in which voltage is low, giving rise to several peaks per constituent [149]. The commonly held notion that the pH gradient becomes stationary with time is an illusion due to the logarithmic scale of pH nomenclature; it is only when one compresses the scale over 8 orders of magnitude logarithmically, that pH gradients appear more stable than the patterns of carrier ampholytes from which they are produced (Fig.36).

Since the pH gradient is always changing, the protein in electrofocusing is characterized not by its position on the gel (R_f) as in PAGE, but rather by the pH at which it is found. That fact makes it neccessary to determine the pH gradient on the gel prior to staining or assay of protein.

The simplest way to determine the pH gradient is to cut the gel into half *manually* by razor blade, then to subdivide the halves until 16 slices are obtained. The halving can be done by eye with a little skill and experience. The slices are suspended in 0.5 mL 0.025 M KCl and analyzed for pH after one hour or two of diffusion. The protein staining pattern in this case is obtained from a duplicate gel. This introduces some inaccuracy, as does the correction of migration distance, measured on the shrunk or swollen stained gel, to that on the unstained gel on which pH was determined.

A better way to measure pH on gels is therefore by *contact* pH *electrode* (see Section 4.11.10). These electrodes still have at present the disadvantage of great fragility, although it is possible to construct sturdy metallic electrodes for that purpose which have the further advantage of being applicable in the cold [150]. Since the response time of contact electrodes, as of any others, decreases with time of use, it is better to use them manually, waiting for a constant pH value at each measurement, than to pass a contact electrode across the gel at constant speed, using a mechanical pH measurement device (Chapter 4). The manual approach has the further advantage that it is not influenced by the pH dependent swelling of the gel and its consequent irregular shape. The electrode can be manually set onto the gel surface equal-

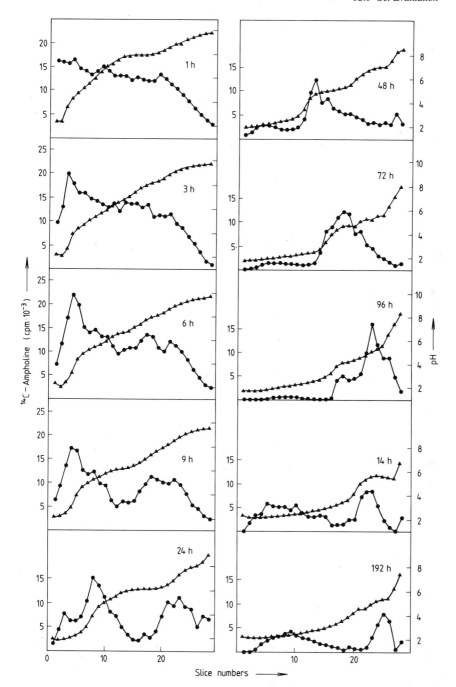

Figure 36. Dynamics of pH gradients (▲–▲) and ^{14}C-carrier ampholytes (●–●) (pI range 3–10 SCAMs), as a function of the time of EF (200 V/5 cm gel, 10 %T, 2 %C_{Bis}, 0 °C. Data of [142].

ly well at constricted or swollen points on the gel, while a mechanical device that follows an irregular surface has not been constructed to date. A suitable stand for the electrode, which allows one to pass it across the gel at regular distance intervals, is available (Chapter 4). Usually, a measurement is taken in 5 mm intervals, allowing one to measure the pH gradient on a gel within a few minutes. The gel is then either stained or assayed.

12.8.2   Protein Staining and Assay

12.8.2.1   Staining of proteins

The staining of proteins on electrofocusing gels is the same as in PAGE (Section 7.8.4) if simple buffers are used as carrier constituents. If SCAMs are used, these stain as the proteins in the conventional procedures used in PAGE. The finding originally of Reisner et al. [151] that 3.5% perchlorate fixes proteins without fixing SCAMs has been a solution to this problem, although it appears that 3.5% perchloric acid (PCA) is not as universal, and not as effective a fixative for proteins as 10 or 12.5% TCA (used in PAGE). The most sensitive no-background staining procedure for EF gels available at this time is Procedure B of Vesterberg [152]. The gel is incubated at 60 °C for 30 min in 40 mL of solution A (9 g sulfosalicylic acid, 30 g TCA, 236 mL water). It is then placed into a 10 mL screw cap tube filled with solution B (0.1 g Coomassie Brilliant Blue G-250, 200 mL water, 8.7 mL 61.3% PCA, with stirring to 300 mL, filter), and re-incubated at 60 °C for 30 min. The gel is stored in solution B diluted 1/100.

12.8.2.2   Assay of gel slices

Assay of protein activities on *polyacrylamide* EF gels follows essentially the procedure described for PAGE (Section 10.3.4.4), but is more difficult because the gel is usually soft ("non-restrictive"), elastic (if DATD crosslinked) and unevenly swollen at different pHs and consequently difficult to slice transversely, and because the protein is isoelectrically aggregated or precipitated and consequently reluctant to diffuse out of the gel. The only *slicer* presently capable of handling the transverse slicing problems of EF gels is the electrovibrator slicer (Section 4.11.3) if its vibration frequency is sufficiently lowered by use of a rheostat. The degree of lowering has to be individually determined for the type of EF gel used, since its consistency not only varies with %T and %C but also depends on the type and concentration of carrier constituents in the gel (Chapter 5). Additional aids in using this slicer are the embedding of the gel within its groove in 1 to 2% agarose, or the construction of a slotted Plexiglas frame into the slots of which the cutting wires fit; this allows for cutting through the gel beyond its lower surface [57].

Granular gels are extruded from the gel tube by freezing the gels containing 25% glycerol in liquid nitrogen, peeling of the cracked glass tube and/or pushing out of the gel with a Pyrex rod after sufficient warming. The gels are then sliced by razor blade [54]. Possibly, the Mickle slicer (Section 4.11.3) might also be applicable to frozen granular gels.

To diffuse the proteins out of the gel for assay, it may be necessary to *solubilize* the isoelectric proteins at high (10.5) or low pH for short times (5 min) with very thorough temperature control at 0 °C (ice-water bath), prior to neutralization and assay. Obviously, some activities may not survive that rather harsh treatment.

12.9 Preparative EF Procedure

12.9.1 On Sucrose Density Gradients

We have already pointed out that the original procedure of EF on sucrose density gradients is milligram-preparative, but that it is burdened by two inherent defects: Sedimentation of isoelectric proteins and, relative to gels, a larger degree of distribution overlap. A third defect is that in the commercial models of this type, zone mixing occurs at the single point of elution, and that the electric field is cut off prior to elution, giving rise to diffusion spreading. These latter defects can however be remedied [153]. To apply the method to weakly acidic and basic anolytes and catholytes, an enlargement of buffer chambers would be necessary. From the viewpoint of ease of recovery, this method remains attractive. Procedural detail is provided by the LKB instruction manual (as an example, see [154]).

12.9.2 On Polyacrylamide Gel

The most practical mode of preparative electrofocusing on polyacrylamide gel consists of excision of the gel slice of interest from one or several gels of 18 mm diameter, electrophoretic extraction and concentration of the protein by MBE, and purification of the extract by gel chromatography, as described for PAGE (Section 14.1) (Fig.37) [155],[64],[65]. The procedure differs from that applied to PAGE only by the need to solubilize and disaggregate the protein in the gel slices prior to MBE. This is done by a short exposure to either basic or acid pH as described in Section 12.8.2 above.

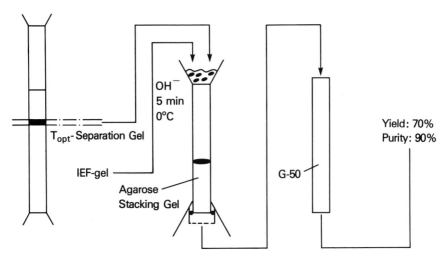

Figure 37. Schematic representation of a method for protein isolation by slicing of PAGE or EF gels, slice extraction and protein concentration by MBE with collection of the protein in a small volume cup, and purification of the cup contents from non-proteinaceous contaminants by gel filtration [64],[65].

12.9.3   On Granulated Gels

An EF procedure for 18 mm diameter Sephadex containing gel tubes has been described [56],[99],[120],[54], including details on gel slicing. For a preparative application, the protein may be allowed to diffuse from the slices, or one may wish to collect the protein by electrophoretic extraction as described in the section above. For a procedure of preparative gel EF on granulated gels, using horizontal slab apparatus, see [33],[34],[100].

12.9.4   By Constituent Displacement

In principle, the displacement of carrier constituents from the gel by choice of an anolyte higher in pH than the terminal anodic pH of the pH gradient can be used to displace and collect the protein [121],[144]. But the rates of displacement appear to be slow, and solutions of practical isolation problems with evidence of good yield and purity via this mechanism still remain to be demonstrated.

12.9.5   SCAM Removal

If SCAMs have been used in EF, their removal from the isolated protein represents a generic problem for all of these preparative approaches. The easiest solution to that problem is to pass the original stock solution of SCAMs prior to EF over an ultrafilter which retains species in excess of 1000 MW [e.g. an Amicon Corp. (182 Conant St., Danvers MA 01923) Diaflo membrane UM2] [156]. The small molecular weight SCAMs can then be removed by dialysis or gel chromatography.

A more risky approach is to separate the SCAMs from the isolated protein by ion exchange chromatography, taking advantage of the fact that the charge densities of SCAMs are higher than those of proteins. This allows them to adhere to 8-10% crosslinked Dowex 50 while proteins are not bound and consequently eluted [157]. The risk in this procedure is, however, that some or all of the protein is lost by adsorption to the backbone of the resin. Vacuum dialysis or electrodialysis on a small-volume apparatus (Micro-ProDiCon, Pierce) has been successfully applied to bulk removal of SCAMs from proteins in SDS-containing buffers [158]. This approach, like ultrafiltration on a hydrophobic Amicon membrane, risks losses through membrane adsorption, although the risk of adsorption on a hydrophilic membrane may be less.

13   Preparative Protein Extraction
     from Gel Slices

Here, as anywhere else in the book, the term "preparative" will be applied exclusively to the milligram-preparative or gram-preparative scale per single protein component.

Compared with the various chromatographic tools for size fractionation or charge fractionation, preparative gel electrophoresis has a bad name in biochemistry. Most workers feel that it is a risky procedure, leading often to partial or total loss of the protein of interest, and that it is too laborious and artful. This opinion frequently rests on negative experiences either with the diffusion of protein from gel slices, or with large-scale elution-PAGE apparatus. However, there exist also real *problems* which support the widespread distrust in preparative electrophoretic methods. Resolution is of course the same as at the analytical level, but the detection method differs: Instead of merely visualizing the maxima of concentration distributions (the bands or zones), detection methods at the preparative scale reveal the entire distributions. These are invariably found to be grossly overlapping. This is not necessarily a fault of the apparatus design, elution flow rates etc. The same overlapping distributions are detected on analytical gels when activity or radioactivity are being monitored. The overlap shows that gel electrophoresis under presently available conditions is a relatively poorly resolving separation tool; when resolving power is expressed in terms of the numbers of theoretical plates [70], numbers in the hundreds are usually revealed. This is to be contrasted with both chromatographic size and charge fractionations using 5–10 micron bead sizes (HPLC), conducted at sufficiently low elution flow rates, where resolving power can be in the thousands of theoretical plates. The roots of excessive band width in gel electrophoresis remain largely unknown: a) Zone *skewing* due to the different net mobilities under the leading and trailing limbs of the distributions may be involved [162]; b) microheterogeneity due to net charge and/or conformation may be involved, since SDS-derivatization shrinks the width of the distributions [30]. Although the same effect would be expected in chromatography, charge heterogeneity there is undetected in gel chromatographic columns, while size heterogeneity is undetected in ion exchange columns. In electrophoresis, they are combined and may therefore be more apparent. c) Finally, the excessive bandwidth in gel electrophoresis may be an artifact of polyacrylamide and due to its oxidizing power (see Chapter 5). This is indicated by the fact that bands broaden upon rerun of gel slices. A test for that hypothesis has become available through the finding that agarose gels with pore sizes equivalent to polyacrylamide can be prepared [21]; to date, however, this test has not been carried out as yet. Nonetheless, preparative PAGE under the conditions of pH and of pore size optimized at the analytical scale has its specific area of usefulness at this time and is feasible, requiring some but not very much novel technique beyond that used at the analytical scale. The scaling up process is quite unproblematic. Preparative gel electrophoresis has a great *advantage* over isolation by chromatographic methods when the exact conditions for optimal resolution have already been worked out at the analytical scale and merely are to be applied at a proportionally larger scale for preparative purposes. Gel electrophoretic isolation methods if applied under optimized conditions also have advantages with regard to resolving power over conventional chromatographic methods since the fraction of time in which the molecules diffuse (migrate) into and out of pores is 100% in a non-flow medium such as a gel, while during chromatographic flow diffusion into and out of pores is limited to a mere fraction of the elution time. Furthermore, the pore size is tailored to the molecule, and is not merely an approximation to a particular size class. Finally, charge and size fractionation can be effected simultaneously in PAGE (see Chapter 1).

While gel electrophoresis is a sufficient method responsive to both size and net charge differences among species in the sense that it preempts the need for alternative size and charge methods, it may *not* be *sufficient* to isolate a protein from a multi-component mixture. As we have seen (Chapter 11), the optimally resolving pore size can only be defined for protein systems consisting of a few components. Thus, size and charge fractionation by chromatographic methods may be required prior to gel electrophoresis if the mixture is of a complex nature and if a %T optimization is desired. The requisite chromatographic pre-fractionation aiming at a simplified starting mixture for gel electrophoresis does not require any more resolving power than these methods possess under conventional conditions. The alternative, when trying to isolate one protein from a multi-component mixture, is to neglect pore size optimization and to fractionate in a gel of median pore size, as the 7.5% polyacrylamide gel in the early serum fractionations of Ornstein [9]. In particular, slicing out of the protein of interest from such a gel is appealing since this isolation method is quite independent of all protein zones on the gel other than those in the neighborhood of the band of interest. Furthermore, this approach allows for the simultaneous isolation of several proteins from the same gel.

We assume here that the procedure for determining the optimally resolving pore size for the particular isolation problem has revealed a finite %T value and has therefore directed the fractionation path either into preparative PAGE (Section 13.2 and Chapter 14) or into preparative MBE (Chapter 13) or EF (Section 12.9) (Fig.28).

13.1   Preparative Moving Boundary Electrophoresis on Gels

Preparative MBE is the method of choice if the pore size optimization (Chapter 11) has provided objective evidence that the protein of interest is most effectively separated from its neighbors at 0 %T. Although preparative electrofocusing (Section 12.9) is a legitimate alternative to MBE in this respect, gel MBE has the *advantage* that it excels in load capacity over gel EF by one order of magnitude (or over

PAGE by 2 orders of magnitude). Also, MBE is not burdened in the recovery of proteins by the relative insolubility and aggregation associated with the isoelectric state of the protein in EF. Finally, gel MBE is of advantage to those who have followed the fractionation scheme outlined in this book, starting with the experimental selection and optimization of stacking conditions for the protein of interest (Chapters 7 and 8): Remembering, that preparative MBE and analytical stacking are the same process, viewed in the first case from within the system of sequential moving boundaries, in the second from its outside (Chapter 3), it is clear that the conditions for preparative MBE have already been *worked out*, and need only be somewhat refined and scaled up for preparative MBE.

Preparative MBE is most suitable for the fractionation of a *small number* of charge isomeric proteins, or those with $T_{opt} = 0$. If the number of components to be separated by alignment in order of mobility at the steady-state becomes excessive, it takes a correspondingly long time to reach the steady-state. This time is experimentally limited by the fact that in MBE, the stack moves across the gel and then runs off into the lower buffer reservoir. To gain the necessary time, one may extend the gel length to about 50 cm. But any further extension becomes unwieldy and impractical. For exactly the same reason, multicomponent "spacers", i.e. small MW species with mobilities intermediate between the proteins [159] even if they existed would be inapplicable to preparative gel MBE [69].

The same limitation holds with regard to protein *load*: The higher the load, the longer it takes to establish the steady-state. Since experimentally, the passage of a stack through a 50 cm long tube at a tolerable voltage (2 V/cm only for reasons indicated below) takes 2–3 days, the load is limited to about 60 mg/cm^2 of gel for a protein system of a few components [69],[160]. This "limitation" of course is by orders of magnitude less than the load limits of other methods.

As pointed out already, there is a serious limitation of preparative gel MBE with regard to *voltage*. This is due to the fact that the stack of proteins, at concentrations in the range of 20 to 100 mg/mL within the stack [7],[69], is highly resistive in the electric field. The high local voltage across the stack leads to relatively much Joule heat and heat denaturation of the protein, unless the total voltage across the gel is severely reduced. In a representative case, voltage across an 18 mm diameter gel of 50 cm length is limited to 75–100 V, which translates into running times for preparative gel MBE of 2–3 days [160].

The *procedure* of preparative gel MBE must be preceded by 3 analytical-scale and one scaling-up experiment:

- a refinement of the trailing ion mobility,
- determination of maximal load at a low voltage,
- determination of maximal voltage at maximal load.

This is followed by

- a scaling up of gel length, gel diameter and protein load.

13.1.1   Refinement of the Trailing Ion Mobility

Resolution between charge isomeric species in gel MBE depends critically on the specificity of the range between trailing and leading ion mobilities for the proteins that are to be resolved. Another way of viewing this requirement is in terms of the pH gradient across the stack. Thus, a rather flat pH gradient across the stack seems required to achieve good resolution in MBE [160]. Experimentally, this means first of all that the protein of interest be stacked in the selected subsystem, but unstacked in the subsystem higher in trailing ion mobilities by one step in the order of column 2 of the Buffer Systems Tables (Fig.4 of Appendix 1) or the Subsystems Table of the Extensive Buffer Systems Output (column 1 of Fig.5 of Appendix 2). Since the "steps" along the columns of the tables are in arbitrary intervals, it appears commendable to even interpolate between the two crucial steps, that at which the protein of interest is stacked, and that at which it is unstacked. Since the step size between trailing ion mobilities of the subsystems is even greater in the Buffer Tables (column 2 of Fig.4 or Appendix 1) than in the extensive output, intrapolation becomes even more important when the tables are used. The procedure of selecting trailing ion mobilities, and the procedure of running stacking gels in the various gel buffers to determine, whether or not the protein of interest is stacked in any of them, is exactly identical to that described in Chapters 7 and 8.

13.1.2   Determination of Maximal Load at a Low Voltage

Once the most narrowly defined trailing ion mobility is selected, the maximal protein load is found which can be tolerated within the limits of migration distance (time) and voltage. Thus, a series of stacking gels of 50 cm length are run at very low voltage (e.g. 75 V) in the tube apparatus with elongated lower buffer reservoir (60 cm length for Part C of Fig.5). Gels are loaded with amounts of oligocomponent protein mixture increasing from 2 mg to 100 mg. In each case, the formation and maintenance of hair-sharp zone boundaries across, and the width of, the "extended stack" is followed. In this experiment, the stack is usually visible as a yellow-brown zone of protein (at the high protein concentrations obtaining within the stack). The front edge of the stack can frequently be marked by a tracking dye; the trailing edge can usually be detected as either a sharp yellowish zone or as a refractile zone of protein or of protein which has bound some dye component. The maximal protein load compatible with maintenance of sharp zone boundaries and constant stack width within the given migration distance of 50 cm is selected.

As the protein load in gel MBE is increased, the stack appears to assume an increasingly swollen appearance, extending the gel diameter as it migrates through the gel, in the manner in which passage of an oversized prey can be followed in a snake. Therefore, in preparative gel MBE even more than in analytical scale MBE, it is essential to select as elastic a gel as the 5 %T, 15 %C_{DATD} gel. The elasticity of such a gel is sufficient to contract the gel diameter to its original value after passage of the stack.

13.1.3 Determination of Maximal Voltage at Maximal Load

The only remaining analytical experiment attempts to arrive at the maximal voltage compatible with the maximally tolerable protein load. This voltage is of course limited by the amount of Joule heating at the center of the gel, and can therefore be ascertained from analytical scale experimentation only as a first approximation. For this experiment, only a single 6 mm gel of 50 cm length is needed. After the extended stack has formed and exhibits constant width, indicating the attainment of the steady-state, the voltage is stepwise increased every few cm of migration, until a white protein heat precipitate starts to appear in the stack. A voltage value 20 or 30 V below that giving rise to the heat precipitate is selected.

13.1.4 Scaling up of Gel Length, Gel Diameter and Protein Load

The scaling up to a preparative load level is done in proportion to the cross-sectional area of gel.

If the desired load capacity is 2–10 mg (10 to 50-fold that of on a single analytical scale gel in PAGE), the tube diameter of the gels of 50 cm length is maintained at 6 mm. If the desired load capacity is 10 times larger (20–200 mg), tubes with 10-fold cross-sectional area are selected, i.e. tubes of 18 mm diameter, keeping all other conditions constant. The upper buffer reservoir with 6 grommets for 18 mm ID tubes is then used with as many of these tubes as needed. Thus, six tubes of 2.54 cm^2 contained in a single apparatus provide a load capacity of 120–1200 mg of an oligocomponent mixture of proteins.

The maximal voltage tolerated on 6 mm gels (Section 13.1.3) is applied with care to immediately detect any white protein heat precipitate appearing in the stack. If it occurs, voltage is reduced to an acceptable level. Usually, the heat precipitate is reversible if detected early enough.

The attainment of the steady-state is again recognized as a first approximation by the constancy of stack width. After the stack has migrated a few cm at the steady-state, usually close to the bottom of the tube, electrophoresis is discontinued. The tube is circumferentially scratched by triangular file 1 to 2 cm below and above the extended stack. A glowing Pyrex rod (O_2 flame) is touched to the scratch, followed by a stream of water from a wash bottle, to crack the tube. The portion of the tube containing the extended stack is then broken off manually, and the gel is extruded by reaming as described in Section 7.7. The gel is then sliced into 1–2 mm slices using the transverse wire slicer [59] scaled up from 6 to 18 mm gel diameter and from 5 to 10 cm gel length (Fig.6 of Appendix 4). Gel slices are suspended in 0.5 to 1.0 mL catholyte of the MBE buffer system used for extraction and concentration. This buffer system is selected, and extraction, concentration and purification of the protein is carried out as described in Section 13.1.5.

13.1.5   Slice Extraction and Protein Concentration

Protein extraction from gel slices with simultaneous concentration of the protein (Fig.37) is the preferred isolation method because it requires no separation methods other than those that already have been worked out at the analytical scale. Furthermore, the risk of losing the protein is minimal in this method. The procedure consists of

- gel electrophoresis,
- non-destructive protein detection,
- gel slicing,
- extraction and concentration of the protein by MBE
  followed by collection,
- purification of the collected protein and change to a volatile buffer, using gel chromatography.

13.1.5.1   Gel electrophoresis

Gel electrophoresis is carried out at the optimized pH and pore size in gels of 18 mm ID, using a 10-fold protein load per gel as compared to the analytical scale 6 mm ID gels. Conditions are exactly those used at the analytical scale, except that the upper buffer reservoir for six 18 mm gel tubes (Part B2, Fig.5) is used. Gel volumes and current are ten-fold that used analytically per tube. However, a strict proportionality of the current assumes that Joule heat dissipation is equally effective on 18 mm as on 6 mm gels. To the degree that this assumption does not hold one may have to reduce the current somewhat below that calculated on the basis of the ratios in gel surface area. Such reduction may not be necessary if care is taken to provide efficient enough heat transfer by a sufficient rate of stirring of the lower buffer and by using high enough coolant flow rates.

13.1.5.2   Non-destructive protein detection

Non-destructive protein detection on the gels to allow one to detect and excise the zone of the protein of interest, can be accomplished a) by transverse slicing of each gel and direct analysis of slice eluates; b) by staining a parallel gel; c) by longitudinal slicing of a guidestrip which is then stained or assayed; d) by prestaining of the protein; e) by the natural fluorescence of proteins at sub-freezing temperatures.

a) Gels are sliced transversely by the gel slicer type most suited for the particular gel concentration used and available for wide-diameter preparative gels (Section 4.11). Each slice is suspended in upper buffer of the analytical MBE system, and aliquots are analyzed by PAGE on a multichannel gel slab while the remaining slices and diffusates are stored in the cold. The protein of interest is found by staining or assay of the slab gel [160]. This method of direct analysis of all gels has the advantage of avoiding assumptions or calculations, but is also rather laborious. Potentially, it could be made much less laborious by a preparative slab technique. But its development will depend on the availability of suitable gel slicers for thick slabs [196] which are not as yet commercially available at this time.

b) Location of a protein zone by staining of one gel out of several unstained gels is wasteful of protein and requires strictly identical gel volumes, sample volumes and geometric conditions in each, but it has the advantage of not requiring special apparatus or techniques. It is important, however, to correct the characteristic R_f value of the stained protein for the change in gel length concomitant with fixation and staining, before applying it to the unstained gels for locating the zone of interest.

c) A longitudinal slice can be obtained either manually, using a microtome knife blade, or using a longitudinal gel slicer (Fig. 7 of Appendix 4). In either case is it desirable to sacrifice as little of the gel as possible, and therefore to obtain as shallow a longitudinal slice as can be comfortably sliced off. Usually about 1/2 of the radius of the gel cylinder is the point of application for the blade or wire. Since a guidestrip can be obtained from each gel, it avoids the assumption of equivalence of several gels. When determining the R_f of interest, it is again neccessary to correct for change of gel length on staining (see b). The disadvantage of that method is that longitudinal gel slicers work best with mechanically stable gels, and least well with elastic DATD crosslinked gels. Manual slicing often gives uneven longitudinal gel slices.

d) It is possible to stain the protein sample prior to electrophoresis, either noncovalently with a small amount of Coomassie Blue [163], or covalently with amphoteric Drimarene or cationic Remazol dyes [164]. However, all these reactions involve usually denaturing conditions, and are therefore primarily of value to separations of proteins without activity.

e) Protein location by the natural fluorescence of the tryptophan residues of proteins at very low temperature appears to be a promising detection device at relatively high protein concentrations in the zone [165].

13.1.5.3   Gel slicing

Once the protein of interest is located on the unstained 18 mm gel(s), they are sliced transversely if that has not been already done during the location procedure. The gel slicer types available for wide-diameter preparative gels are listed and their specific areas of application are pinpointed in Section 4.11.3. The gel slices of about 1 mm thickness carrying the protein of interest are suspended in the upper buffer of the MBE system selected for electrophoretic extraction and concentration (see Section 13.1.5.4 below). The buffer should also contain 10–20% sucrose and tracking dye. If several preparative gels have been run, slices carrying the protein of interest are pooled.

13.1.5.4   Extraction and concentration of the protein by MBE followed by collection

The protein is to be extracted by MBE from the gel slice, concentrated into a stack and collected in a small volume chamber with dialysis-membrane-floor attached to the gel tube.

Choice of MBE buffer system

For electrophoretic extraction and concentration, one wants the most rapid protein migration rate compatible with the maintenance of native properties. Also, selective stacking of the protein of interest is no longer required, which makes it unnecessary to electrophorese close to the pI of the protein. Thus, a MBE buffer system operative at pH 10.5 (No.1, Appendix 1) appears useful in this application if protein activity can be maintained at such a high pH; correspondingly low pHs (systems 15,16 of Appendix 1) appear equally applicable, although in the acidic pH range voltage gradients are relatively diminished due to proton migration. Leading and trailing ion mobilities should be selected to include within their range the mobility of a tracking dye; this is not a problem at extremes of pH where bromphenol-blue or -red mark the stack in alkali, methylgreen, brilliant green or toluidine blue in acid.

It is of course not necessary to select a new stacking system for the extraction and concentration step. It is certainly more convenient to apply also at this step the MBE system optimized for PAGE fractionation (Chapters 7, 8). But one pays some price for that choice in terms of the time of extraction which may be inconveniently long [155].

Apparatus

The apparatus required for protein extraction and concentration by MBE is a collection chamber for the stack at the end of a gel tube. One can use the ordinary tube apparatus, e.g. with 6 or 18 mm ID tubes, and tie a little dialysis bag to the bottom of the tube. Or, one may construct tubes with a permanently assembled collection chamber [65]. The design of the collection chamber has been steadily improved to the model shown in Fig.20. This model consists of a minor accessory – the collection cup – to the general tube apparatus for PAGE (Fig.5). A funnel shaped tube reservoir which holds the gel slices is used for convenience. It can be made in a variety of sizes, depending on the desired number and volume of gel slices that are to be extracted. At least 4 "funnels", i.e. 18 mm ID tubes topped by a tube section of enlarged diameter, fit into the ordinary upper buffer reservoir. These "funnels" can be entirely filled with gel slices, since the surrounding upper buffer reservoir holds the common upper buffer. Alternatively, one can use an elongated lower buffer reservoir with 18 mm ID tubes long enough to hold the desired volume of gel slices.

Gel

Since protein migration is to be enhanced at this step, molecular sieving effects should be reduced as much as possible. For large proteins, this makes it desirable to conduct MBE on 1% agarose gels. This has also advantages over polyacrylamide with regard to the purity of the product (see Section 13.1.5.5). Unfortunately, electroendosmosis-free agarose (Iso-Gel, Marine Colloids) interferes with MBE in buffer systems of positive polarity [97], so that it cannot be considered generally applicable to the extraction and concentration step at this time.

In those cases, where 1% agarose is inapplicable, 5 %T, 15 %C_{DATD} appears as the next best choice. For small molecular species, a "non-restrictive" low %C_{Bis} crosslinked gel may be preferable, since at sufficiently high %T values it combines mechanical stability with high polymerization efficiency and average chain length (Chapter 5).

Gel volume should be sufficient to allow enough time for extraction (we recommend 4–5 sample volumes in the procedure below), but it also should not exceed a volume which can be saturated with regard to protein adsorption to polyacrylamide [50] without appreciable sacrifice in yield. Thus, if microgram amounts only are available, the extraction and concentration by MBE should be carried out on tubes or 6 mm ID or less.

Procedure

The tube to be used for MBE is washed, neutralized and coated with 1% agarose (Iso-Gel) as described in Section 4.5.3. If polyacrylamide is being used at low %T (6 %T or below), the tube is coated with Gelamide 250 (see ibid.). Since it is not necessary to remove these gels intact from the tube, polyacrylamide gels may be formed in tubes coated with Silane A-174 (see Section 4.5.3).

The tube is filled with stacking gel of the desired MBE buffer system and gel concentration, i.e. usually either 1% agarose (Iso-Gel) or 5 %T, 15 %C_{DATD} are used. The gel volume should be at least 4 to 5-fold the sample volume. To improve mechanical stability, funnel tubes are filled with gel to just reach the full diameter of the funnel. Agarose gels are poured at 60 °C, left to gel at room temperature overnight and are cooled and used for gel electrophoresis the next morning.

The sample consisting of gel slices, upper buffer, 10–20% sucrose, and tracking dye is loaded onto the gel. The slices should be fully covered by the buffer. Then upper buffer is overlayered sufficiently slowly so as not to disturb the interface with the sample. This can be done manually by syringe or pipet, or by mechanical pump. To minimize diffusion of the sample into the upper buffer, the current may be turned on once the upper buffer level over the sample reaches 2–3 cm; overlayering of the upper buffer is continued with the current turned on until the upper buffer reservoir is filled.

Electrophoresis is conducted at 6–7 mA/cm^2 of gel until the stack, visualized by the tracking dye, reaches near the bottom of the gel tube. The collection cup (Fig.20) filled with 1:4 diluted gel buffer is then attached, and electrophoresis is continued at 3.0–3.5 mA/cm^2. After entry of the stack into the elution cup, electrophoresis is continued for 2 to 3 h, before the contents of the cup are collected. Usually, the entire MBE experiment requires one entire workday of 8–10 h. Of course, it could be run at half-speed overnight, but in general, the time between slice excision and lyophilization of the purified extract should be minimized to maintain the native properties of the protein, so that maximal currents are advisable. Also, in non-routine isolations, one does not want to risk electrophoresis without supervision with the elution cup unattached as necessary to maintain a maximal migration rate. After collection, the cup may be refilled with upper buffer, and electrophoresis be continued overnight to collect a second crop.

13.1.5.5   Purification of the collected protein by gel chromatography

The contents of the collection cup are heavily contaminated with polyacrylamide-like migrating materials when MBE is carried out on polyacrylamide gels. The degree of contamination with non-proteinaceous impurities is negligible, when agarose is used as the medium for MBE. To date, the method used to purify the protein has been gel chromatography on Sephadex G-50 in cases where the protein elutes from these gels are relatively undiluted with the void volume of the column [155]. Column diameter of 0.9 cm, column lengths of 50-60 cm, flow rates of 0.5 mL/min, 0–4 °C, Sephadex purified by extraction with water at the boiling point for 24 h in a Soxhlet apparatus (Appendix I of [112]) and 0.02 M NH_4HCO_3 made from NH_4OH and bottled CO_2 have been used for that purpose. The protein emerging with the void volume of the column is lyophilized. The long-chain impurities are retained on the column. Columns may be used a few times. Optimally, the chromatography step is carried out on the same workday as the extraction by MBE, requiring a total of approximately 14 h. The chromatographic step should increase the purity of the protein to 90% or more. The recovery of protein through steps 13.1.5.4 and 13.1.5.5 has been nearly that high.

13.2   Preparative Resolving Gels (PAGE)

Although a great many forms of preparative PAGE have been reported in the literature [66], only those two will be considered here which appear to be simple prototypes of many of the other methods. The choice between the two rests on the amount of protein which is available and which is to be isolated. If a few milligrams of isolated protein are required as in most cases, the method of choice is a scaled up analytical gel, excision of the gel slice of interest followed by electrophoretic extraction and concentration of the protein from the gel (this chapter). If the protein requirement is large, the more difficult procedure of elution-PAGE is needed (Chapter 14).

Procedurally, protein preparation by extraction from the gel slices of resolving gels follows Section 13.1.5 with the sole exception that resolving gels at the optimally resolving gel concentration (Chapter 11) are sliced and that the gel slice of interest is defined by its R_f (defined here as slice number of the protein of interest divided by the slice number of the moving boundary front).

The load capacity of PAGE is one to two orders of magnitude less than that of MBE, *viz.* of the order of 0.1 mg of a single protein component/cm^2 of gel. Keeping in mind, that the maximal gel thickness compatible with PAGE at a practical rate in gels immersed in thermostated buffer is 18 mm, the cross-sectional gel area needed for a particular preparative resolving gel can be calculated. Six cylindrical gels of 18 mm, with a gel surface area of $6 \times 2.54 = 15\ cm^2$ have a load capacity equivalent to that obtainable by elution-PAGE apparatus (Chapter 14). Gel slicers for these cylindrical gels are available (Section 4.11), making the isolation of 1-2 mg of protein by slice extraction from up to 6 cylindrical 18 mm diameter gels practical [158] although at first sight such approach appears clumsy compared to the sectioning of a single 18 mm gel slab of appropriate length. The problem, as pointed out in Chapter 4, is that such slab apparatus with adequate temperature control is not available; nor has a suitable gel slicer been designed to date.

The extraction of the protein from resolving gel slices needs to be prolonged in proportion to the restrictiveness of the gel concentration for the protein of interest. Thus, the MBE gel should be longer than that used for the extraction of preparative MBE gel slices (Section 13.1).

In a representative case [158], preparative resolving gels have yielded recoveries of 60–70%. Considering that the procedure runs little risk in losing the protein, resolving gel slice extraction by MBE appears as the method of choice for the isolation of 1–2 mg of protein [64]. It shares, of course, with all of the other gel electrophoretic separations the problem of distribution overlap (see above). This makes it unlikely that a homogeneous protein can ever be isolated in a single pass of the procedure. This fact forces one into 2-stage fractionations and the need to sacrifice yield to homogeneity within the reasonable limits posed by the law of diminishing returns [155],[158],[166].

14 Preparative Elution-PAGE and -MBE

We designate as elution-PAGE that mode of preparative gel electrophoresis in which the protein is allowed to migrate across the entire resolving gel and into an elution chamber with a membranous floor, from which it is continuously or discontinuously displaced by buffer flow [66]. The present advantage of that method over gel slicing and slice extraction (Section 14.1) relates to load capacity. In elution-PAGE, apparatus with 20 cm^2 gel surface can be constructed which is capable of sufficient heat transfer to allow for a fractionation time of about 5 h. This is done by polymerizing an annular gel which is capable of being cooled from both inside and outside [167]. Such a gel, however, cannot be sliced with presently available tools. An extended gel slab with equivalent gel surface area could be sliced with a large knife and protein could be then extracted as described in Chapter 13. But no more accurate and less artful slicing method for such a preparative gel slab is presently available. A cylindrical gel of equivalent surface area would be easier to slice but would have to be run extremely slowly, probably for days, in view of the poor Joule heat transfer from the interior of a fat cylinder to its cooled surface. Thus, elution-PAGE with 20 cm^2 of gel remains relevant for gel electrophoresis, in spite of its requirements of apparatus and procedural skills, and after the widespread disappointment with such apparatus that we have referred to in the introduction to Chapter 13.

14.1 Apparatus

Since the gel slice extraction method discussed in Chapter 13 is much easier and safer a procedure than elution-PAGE, only those large elution-PAGE apparatus types should be considered which have a load capacity unobtainable at present by gel slicing methods. Those are the ones with more than 10 cm^2 of gel surface area [66]. We will also restrict ourselves to the prevalent design type among commercially available apparatus, which also happens to be the original design [167] and the one we have experience with [50],[67],[68],[69],[112]. This design type (Fig.21) consists of an internally and externally cooled gel cylinder of annular cross-section, connected to an upper and lower buffer reservoir. The width of the annulus is 15 to 18 mm for the same reason that preparative gel tubes do not exceed that diameter, i.e. because Joule heat cannot be dissipated efficiently enough at practical rates of PAGE if the diameter were any wider.

The sole design element different from, and additional to, ordinary tube apparatus is a 1 mm thin elution chamber at the bottom of the gel cylinder. Like the collection chamber described in Chapter 13, this elution chamber has a membranous floor. Buffer sweeps the elution chamber either continuously or discontinuously, and thus washes the protein components which have migrated into the elution chamber into a collection system. The sweeping of the elution chamber may be upwards or downwards [17] through a central capillary, or, less efficiently for large diameter cylinders, sideways. The nature of the membrane varies among apparatus models. Dialysis membrane has the disadvantage that it swings with the pulsation of the elution pump and therefore easily occludes the 1 mm thin elution chamber. Small pore gels used as the membrane have the disadvantage that their pore size distribution necessarily permits some entry and loss of protein. The most satisfactory membrane seems to be Corning 7930 porous glass, which is rigid and impermeable to proteins. However, it is also relatively impermeable to buffer ions as compared to protons, so that it generates pH imbalances. They are counteracted to some degree by selecting a lower buffer which is 3 to 5 times more concentrated than that in analytical scale PAGE. A commercial apparatus functioning in this manner is the Polyprep-200 apparatus of Buchler Instruments, with 15.8 cm^2 gel surface area.

The mechanical stability problem is aggravated with heavy, large blocks of gel at the preparative scale as compared to the analytical one. Thus, the restriction to Pyrex walls rather than hydrophobic plastics, the need for wall-coating and for hydrostatic equilibration of the gel are relatively exacerbated. In practice, this apparatus needs to be supplemented by a manometer tube, to monitor for the absence of hydrostatic pressure on the gel, by two Mariotte bottles (2 L), by a heavy lab-jack (e.g. Fisher Scientific No. 14-673-10) and a proportioning pump (e.g. AutoAnalyzer Pump, Technicon Instruments Corp., Tarrytown NY 10591).

14.2   Gel

Since the protein in preparative elution-PAGE has to pass the entire gel length, and time dependent diffusion spreading of the zones is therefore aggravated as compared to analytical PAGE, the optimally separating gel concentration, T_{max} rather than the optimally resolving gel concentration, T_{opt} is being applied to elution-PAGE as a first approximation to the real band spreading in gels [43]. For the same reason, relatively short gels are preferable. Since the mechanical stability of large blocks of gel rapidly decreases with gel height, it is necessary to make a practical compromise between low gel volumes providing rapid fractionation and high gel volumes providing the best mechanical stability through adherence to the glass walls of the gel cylinder. Usually, therefore, gel length is not appreciably different from the one used at the analytical scale, with gel volumes ranging from 50 to 90 mL in the Polyprep-200 apparatus.

14.3   Choice of Elution Buffer

Since the elution chamber is very flat, usually of the order of 1 mm, proteins migrate through the chamber relatively rapidly. Once they arrive at the membrane, they are easily bound irreversibly to the membrane, or denatured at the membrane surface. Thus, an elution buffer flow rate needs to be selected which is high enough to counteract the migration to the membrane. Usually, one elution chamber volume per minute is used at the start of protein elution, i.e. with proteins of high R_f. At later electrophoresis times, the elution flow rate may be decreased [68].

Another device for preventing the contact of the protein with the membrane is the deceleration of protein migration in the elution chamber through choice of an appropriate elution buffer pH and ionic strength. The safest of these devices is the increase of the ionic strength of the elution buffer by a factor of 3 to 10. This allows us to maintain the pH of the elution buffer at that of the operative resolving gel (trailing phase). This pH is probably not modified by the Bethe-Toropoff effect [168] due to the membrane, if the buffer flow rate is as high as recommended here. One can, of course, also decrease the pH of the elution buffer relative to the trailing phase (in the anionic polarity, or increase it in the cationic polarity), but then one risks a backward migration of the protein into the gel, or an undesirable piling up of protein at the lower resolving gel surface.

14.4   Procedure of Polymerization

Prior to introducing the polymerization mixture into the apparatus, it is crucially important to *seal* the bottom gel surface, the entrance channels into the elution chamber and the central capillary. The latter is filled with water, prior to its closure by clamp, as a further precaution against entrance of polymerization mixture into the capillary. Also, both the ground glass bottom surface of the central capillary tube and the polypropylene surface of the gel support plate used for polymerization are lightly greased with petroleum jelly.

To be able to correlate analytical and preparative separations, and to be able to predict elution times on the basis of R_f, it is imperative that the conditions of polymerization be the same at the analytical and preparative levels. The sole distinction, as with all wide-diameter gels, is that polymerization *time* for each gel be doubled, from 30 to 60 min.

To provide enough *light* intensity in the Polyprep-200 apparatus, three preparative-scale illuminators need to be aligned concentrically around the apparatus.

To prevent freezing of a thin layer of water on the inner walls of the apparatus prior to introduction of the polymerization mixture into the apparatus, the *coolant* flow should be started only after delivery of the polymerization mixture into the apparatus.

Overlayering is best carried out by proportional pump at flow rates of 1 mL/min or less. It is essential to direct the flow of water to the walls a few mm above the gel, to have cleaned the apparatus walls sufficiently so that the layering water flows down smoothly along the walls and to decrease the flow rate if necessary to such a level, that the gel surface remains quiescent at the point of impact of layering water. Once a few mm of overlayering water have accumulated, the flow is stopped and illumination is begun.

The *withdrawal* of the gel from the polymerization platform, after polymerization is complete, requires a) that the central capillary and the inlets into the elution chamber be opened prior to withdrawal, in order to allow air to enter and to prevent the formation of a negative pressure that would pull on the gel and separate it from the glass walls; b) that the rate of turning the threaded polymerization platform be slow and even enough to allow for immediate pressure equalization through the central capillary and inlet ports into the elution chamber; c) that the apparatus be lifted above the bench surface and held mechanically, using a heavy lab-jack support to which the apparatus is clamped, in order to allow for such a slow turning of the polymerization platform, with visual inspection of the gel and its attachment to the glass walls.

As soon as the polymerization platform is withdrawn, the gel column is *positioned* centrally above the glass membrane (which is built into the top surface of the lower buffer reservoir). The reservoir above the glass membrane is filled with 1:4 diluted resolving gel buffer. The column is slowly lowered into the lower buffer reservoir to a height of 1 mm above the glass membrane, using the lab-jack positioning control. The column may be tilted during insertion to avoid formation of an air bubble under the gel. Excess buffer overflows from the elution chamber. The gel column is then fastened into the lower buffer reservoir (filled with lower buffer) by means of 3 concentric set screws.

It should be noted that in the present Polyprep-200 apparatus, the *height* of the *elution chamber* can be fixed permanently at about 1 mm. This is done, prior to any use of the apparatus, by a 1/2 to 3/4 counterclockwise turn of the ring holding the membrane (Corning 7930 ion permeable glass) against the lower buffer reservoir, starting with zero elution chamber depth.

As soon as the gel column is mounted in the lower buffer reservoir, the gel must be *hydrostatically equilibrated.* This is done by allowing 1:4 diluted resolving gel buffer ("Elution Buffer I") to flow by gravity from the sealed Mariotte bottle (elution buffer reservoir) into the elution chamber. The air outlet tube has been connected to a piece of narrow bore (1/16") Tygon tubing (the "manometer tube") reaching in length to the top of the apparatus and held vertically by a rubber band around the upper buffer reservoir. The central capillary and elution buffer inlet port are opened and the height of the elution buffer reservoir mounted on the lab-jack is increased until buffer flows by gravity into the elution chamber, progressively displacing any air into the central capillary. The capillary is clamped when it is filled with buffer. The entire apparatus is then tilted on the bench, so that the outlet port feeding into the manometer tube is positioned above the buffer inlet port. The outlet port into the manometer tube is opened, and residual air is displaced into the manometer tube. The shrinking air bubble remaining in the elution chamber is followed visually and directed by appropriate tilting of the apparatus into the manometer tube. When the buffer level in the manometer tube has reached that of the gel, the elution buffer inlet port into the elution chamber is closed. The gel column is repositioned on the bench. It is now hydrostatically equilibrated.

The overlayering solution is withdrawn from the gel, by reversal of the direction of pumping through the overlayering tube. The *stacking gel* is then polymerized in identical fashion to that used to form the resolving gel. The gel is re-equilibrated hydrostatically by allowing elution buffer to flow into the elution chamber until the buffer level in the manometer tube reaches that of the stacking gel surface.

14.5   Procedure of Electrophoresis and Collection

The central capillary is unclamped and connected to the pump so that elution buffer is pulled from the elution chamber through the capillary into the pump. Elution *buffer flow* through the central capillary is initiated at 1 to 1.5 mL/min. The upper buffer reservoir is filled with upper buffer with intermittent adjustment of the level of the elution buffer reservoir by the lab-jack positioning control so that the buffer levels in the manometer tube and the upper buffer level above the gel remain approximately the same at all times.

The *sample* is pumped onto the stacking gel surface, using the overlayering tube and pumping rates of 1 mL/min or less. Electrophoresis is conducted at 50 to 90 mA [50]. Eluate collection is started when the stack enters into the resolving gel, to be able to correlate R_fs with eluate fraction numbers.

Upper buffer is *recirculated* at 1 mL/min. Lower buffer is pumped into the lower buffer reservoir at 3 mL/min and allowed to syphon out of the reservoir into a waste receptacle or sink. Both operations use the identical pump as that employed for elution buffer flow, each with the appropriate calibrated tubing sizes (Technicon) to achieve the desired flow rates.

When the moving boundary front marked by a tracking dye enters into the elution chamber, the elution buffer is changed to one equalling the trailing phase in pH, but concentrated by a factor of 3 to 5 (elution buffer II, contained in a 2 L Mariotte bottle, and connected to the outflow line of elution buffer I by a Y-connector so that one can change from one to the other elution buffer by mere reclamping of the buffer outlet lines of the two Mariotte bottles).

The tube number containing the tracking dye which marks the stack, divided by the known R_f of the protein of interest, provides the predicted tube number for the protein peak.

If a continuous elution flow rate deceleration paralleling the R_f [68] is used, the decelerator is activated at the time of entry of the stack into the elution chamber.

14.6   Eluate Concentration

Preparative elution-PAGE with elution buffer flow rates approximately equal to one elution chamber volume per minute necessarily gives rise to rather substantial eluate dilution. To some degree, this can be counteracted by decreasing the elution buffer flow rate in proportion to R_f [68]. In any case, the eluate needs to be concentrated. This is readily done by the identical procedure and apparatus used for gel slice extraction and concentration (Chapter 13). For eluates of 50 to 100 mL, very large funnel sizes are used [161], for smaller volumes, those described in ref. [65] (Chapter 13).

14.7   Computer Simulation

With the (incorrect) assumption that zone spreading in gels equals free diffusion zone spreading, a computer program (DAR001) written by D. Rodbard can be used to predict elution times and resolution in preparative elution-PAGE for any set of 2 components with known $\overline{R}$, K_R, and Y_o values (Fig. 38). This program should facilitate preparative elution-PAGE by allowing for computer simulation of runs under various conditions, once the real degree of zone spreading in a gel of particular %T and %C is measurable and known. To date such measurement has been carried out on a single protein only, using a not generally available apparatus for PAGE with continuous optical scanning during electrophoresis [169]. Neglecting the fact that the protein used in that study was not sufficiently homogeneous, it indicated that log $(D/D_o) = K_R$, where D and D_o are the diffusion coefficients in a gel and the free diffusion coefficients, respectively. This relation conforms to theoretical prediction [70] and should allow one to calculate D at constant temperature, protein load, pH, and ionic strength from band width measurements at various gel concentrations. The simulation of preparative elution-PAGE runs has therefore moved into the realm of real possibilities, although it is not as yet a practical tool as long as no easy general method exists to measure protein band width in gels.

14.8   Preparative Elution-MBE

Assuming that an optimally resolving gel concentration exists at 0 %T, and that the chosen method of separation, for reasons given in Chapter 13, is preparative MBE, elution-PAGE apparatus and procedure as given in the preceding sections

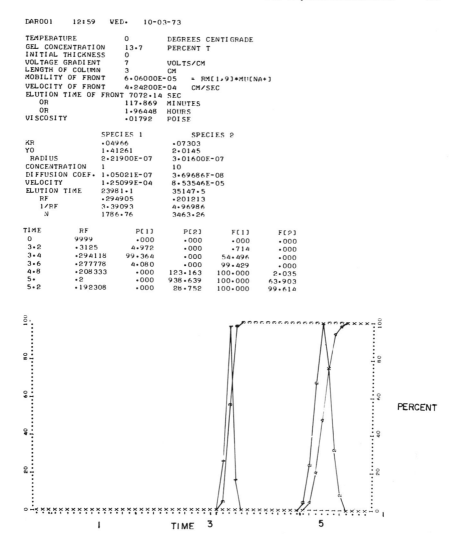

DAR001   12:59   WED.   10-03-73

TEMPERATURE   0   DEGREES CENTIGRADE
GEL CONCENTRATION   13.7   PERCENT T
INITIAL THICKNESS   0
VOLTAGE GRADIENT   7   VOLTS/CM
LENGTH OF COLUMN   3   CM
MOBILITY OF FRONT   6.06000E-05   = RM[1,9]*MU[NA+]
VELOCITY OF FRONT   4.24200E-04   CM/SEC
ELUTION TIME OF FRONT   7072.14 SEC
        OR   117.869 MINUTES
        OR   1.96448 HOURS
VISCOSITY   .01792   POISE

                         SPECIES 1   SPECIES 2
KR   .04966   .07303
YO   1.41261   2.0145
  RADIUS   2.21900E-07   3.01600E-07
CONCENTRATION   1   10
DIFFUSION COEF.   1.05021E-07   3.69686E-08
VELOCITY   1.25099E-04   8.53546E-05
ELUTION TIME   23981.1   35147.5
    RF   .294905   .201213
    1/RF   3.39093   4.96986
    N   1786.76   3463.26

TIME   RF   P[1]   P[2]   F[1]   F[2]
 0   9999   .000   .000   .000   .000
 3.2   .3125   4.972   .000   .714   .000
 3.4   .294118   99.364   .000   54.496   .000
 3.6   .277778   4.080   .000   99.429   .000
 4.8   .208333   .000   123.163   100.000   2.035
 5.   .2   .000   938.639   100.000   63.903
 5.2   .192308   .000   28.752   100.000   99.614

HORIZONTAL RANGE : 0 TO 1 IN INCREMENTS OF .02
VERTICAL RANGE : 0 TO 1000 IN INCREMENTS OF 1

PERCENT

TIME

Figure 38. Input and output of program DAR001 (D. Rodbard) for the prediction of protein elution rates in elution-PAGE. Input parameters for each protein are: K_R, Y_0 and the diffusion coefficient, D', for the particular gel concentration and pH (assumed in the figure to equal the free diffusion coefficient).

can be applied instead of the procedure of protein extraction from gel slices given in Chapter 13. This has the advantage that a single gel of approximately 20 cm² of surface area can be used in lieu of at least 6 cylindrical gels of 18 mm diameter or a gel slab of equivalent area for which neither adequate apparatus or slicing devices exist. The selection of elution gel technique has the disadvantage, compared to slicing methods, of both increased risk of losing the protein during passage through the elution chamber and of eluate dilution.

Procedurally, the sole distinction of elution MBE from elution PAGE is the length of the gel, which in view of the need to reach the steady-state needs to be maximal. In the elution PAGE apparatus under consideration in the previous sections, this is about 20 cm. The load capacity of MBE being 60 mg/cm^2, this small apparatus then becomes a gram-preparative, nearly industrial-scale preparative tool. This load capacity has been demonstrated experimentally on two- to four-component systems [69].

15   SDS-PAGE

To study the protein subunit structure of an isolated protein, or to obtain the composition of a mixture in terms of dissociated and denatured protein subunits (gene products), PAGE is best conducted in the most effectively dissociating detergent, sodium dodecylsulfate (SDS). SDS-PAGE is also needed in the study of water-insoluble, hydrophobic proteins which are insoluble in any of other classes of detergents capable of maintaining native properties (see Chapter 2). Since SDS is able in general to disrupt all non-covalent bonds, SDS-PAGE reveals, in all cases of proteins composed of multiple non-covalently linked chains, not the (post-translational) proteins but their protein-subunits.

SDS having a high CMC (8.2 mM in water [3],[5]) binds to proteins as a monomer under conventional conditions (0.1%). Since the SDS-binding is cooperative, involving both the polar head and the non-polar tail of the detergent, hydrophilic as well as hydrophobic proteins react with SDS, and after exhaustive binding of the detergent assume a constant charge density of the SDS-derivatives corresponding to 1.4 g SDS bound per gram of protein, if a) the protein contains no non-protein moieties, i.e., it is not a conjugated glyco-, lipo -or nucleoprotein, and b) the derivatization with SDS is complete, i.e. the efficacy of reduction and the reaction time at 100 °C have been sufficient.

In the Ferguson plot, the constant charge density of the SDS-derivatives of all proteins gives rise to a common Y_o value for all proteins. In this special case of pure size isomerism, $(R_f)_1/(R_f)_2 = (K_R)_2/(K_R)_1$. Thus, in the specific case of SDS-PAGE where this relation holds, not only is K_R a measure of molecular weight, but also R_f at any one %T value. Accordingly, the molecular weight of denatured protein subunits may be obtained on the basis of R_f, if it is known that the protein of interest does not contain glyco-, lipo- or nucleo-moieties and that it is fully derivatized with SDS.

15.1   Determination of the Adequacy of Derivatization with SDS

15.1.1   Saturation with SDS

The usual derivatization conditions of proteins with SDS, i.e. 5 min boiling in 1% SDS in presence of 5 mM dithiothreitol, may or may not be sufficient for complete disaggregation (e.g. see [105]), dissociation of protein chains, or the complete hydrolysis of disulfide bonds. Thus, the reaction mixture containing protein, SDS and reducing agent needs to be analyzed as a function of time, concentration of SDS, and the intensity of reducing conditions. The latter requires multiple additions of reducing agent to the system, since after addition the reducing agent is rapidly oxidized in air. In practice, prolonged heating of the protein sample in 1% SDS leads to dehydration unless the reaction is carried out in a sealed tube or with a long enough tube to act as an air condenser. (To ascertain that no volume loss occurs during prolonged boiling, the original level should be marked on the tube and water should be added to this mark after the reaction, if necessary.) A sealed tube will also reduce the degree of air oxidation of the reducing agent. Thus, aliquots of the protein sample reacted with various levels of reducing agent and of SDS, and maintained for various reaction times at 100 °C, are analyzed by SDS-PAGE (see Section 15.1.6) to find sufficient conditions for full dissociation, indicated by a gel pattern that does not change further upon increasing the reaction time or concentration of SDS or reducing agent.

Some isoelectric proteins may not derivatize with SDS to saturation level. Thus, proteins with alkaline pI (e.g. RNAse) reacted with SDS at neutral or slightly alkaline pH give species with reduced mobility in SDS-PAGE, rather than the enhanced binding of SDS and mobility expected on grounds of an increased electrostatic (cooperative) interaction with the detergent. Refractoriness of isoelectric proteins to derivatization with SDS would also explain why constant mobilities in SDS-PAGE cannot be achieved below neutrality where most proteins are isoelectric. (The pK of the sulfate groups of about 0 would lead one to expect that SDS-PAGE should be feasible with constant protein mobilities to a pH of about 2 – but this is not found experimentally.) Preliminary experiments in which proteins were derivatized at their pIs have to date failed to substantiate such refractoriness (B. An der Lan, unpublished), but further basic proteins need to be tested. The issue of the SDS-derivatization of isoelectric proteins is of obvious relevance to 2-dimensional EF-SDS-PAGE (see Chapter 16).

2-D PAGE in particular involves the problem of saturation with SDS under reaction conditions with SDS which are much less stringent than those used in SDS-PAGE, i.e. in the absence of heating, reduction and high concentrations of SDS (1%). It appears, however, that a combination of hydrophobic and hydrogen bond breaking agents (such as 2% Triton X-100 and 9 M urea) provides a medium for SDS-derivatization which is equivalent in effect to heating [173],[174]. Other urea-detergent combinations may be preferable; a detailed discussion of this and the non-reduction aspect of "cold derivatization" is given in Chapter 16.

Once the protein is saturated with SDS, conditions need to be found to maintain that saturation during electrophoresis. This is done by finding by systematic SDS-PAGE experiment a SDS concentration in the catholyte sufficient for yielding a mobility of the protein of interest which remains constant upon further increase in the SDS-concentration of the catholyte [78].

15.1.2   Insufficient Disaggregation

Proteins which have a pronounced tendency to aggregate and precipitate at elevated temperatures (a characteristic of surface binding proteins), may exhibit aggregates in SDS-PAGE. The bacterial Filamentous Hemagglutinin (FHA) [105] is an example. The aggregates are recognized by MW analysis and their agreement with MWs in PAGE (in the absence of SDS) for components with horizontally aligned K_R-Y_o-ellipses (indicative of oligomers). It appears that the aggregation is due to heat since larger FHA oligomers (retarded behind the stack) are revealed when the wattage (Joule heat) of PAGE is increased [105]. If the competition between aggregation and SDS-dissociation is shifted by a lowering of the protein concentration, the aggregated SDS-FHA species can be made to dissociate.

15.2   Determination of Non-protein Moieties by Ferguson Plot

Using the sufficient conditions of derivatization with SDS, Ferguson plots are constructed to ascertain whether the plot for the protein subunit of interest intersects the log R_f axis at the same Y_o value as standard proteins which are known not to possess non-protein moieties. The procedures involved in constructing these Ferguson plots are the same as those for native proteins described in Chapter 10. If the protein subunit of interest shares its y-intercept with the standard proteins with statistical significance as evidenced by a horizontal alignment of the joint 95% confidence envelopes of K_R and Y_o on the plot of log Y_o vs K_R, the MW of the unknown may be determined on the basis of R_f (see Sections 15.3 to 15.8). If no such horizontal alignment is found, i.e. charge densities of unknown and protein standards are not equal, MW have been determined by K_R in the same manner as for native proteins (Chapter 10). The relevant output of program GIANTRUN for random coiled species is applied in that case. This approach, however, is also inaccurate, since it assumes the chemical homogeneity of unknown and standards.

15.3   Protein Hydrolysis under the Conditions of SDS-PAGE

When the derivatization with SDS is applied to proteins in the presence of bound fatty acids, hydrolysis of the protein may occur. E.g., the above-mentioned FHA when derivatized with SDS at pH 8 after storage of the protein in formic acid buffers (pH 3.5) hydrolyzes [105]. In this context, it is important to remember that all proteins appear to bind unsaturated fatty acids [110] which are present in lipid-containing samples. Lipoproteins and the fatty acid binding protein, albumin [170], are further examples. The mechanism of the hydrolytic reaction has not been studied to date. It is known that acid catalyzed hydrolysis is vastly accelerated in SDS-proteins [171]. Although that study was limited to acid catalyzed hydrolysis, the important fact is that the microenvironment of the peptide bond is modified by the cooperative binding of SDS, and presumably by the proximity of sulfate, to labilize the peptide bond. The bound fatty acid seems to further enhance this lability, probably by the same mechanism, to allow for either acid or base catalyzed proteolysis at pH 8 in essentially the same way by which enzymatic hydrolysis of the peptide bond takes place at neutral pH. Thus, FHA reacted for 40 min in 1% SDS, 100 °C, if exposed to formic acid (but not acetic acid) prior to reaction hydrolyzes, while without such exposure, its fragmentation occurs after a reaction time of 24 h. There is additional evidence with a serum preparation of plasmin indicating that reducing conditions during the reaction with SDS at pH 8, as well as an alkaline pH of the SDS-gel may contribute to proteolysis (Fig.3 of [172]). This would suggest an addition of (protein)-SH across the double bond of bound fatty acid (derived from the heating step in presence of serum-lipid), a reaction known to proceed at an alkaline pH. But the available data do not suffice to relate such reaction to a proteolytic mechanism.

15.4   Apparatus

The apparatus requirements for SDS-PAGE are two-fold. First, to determine saturation of the protein with SDS on the basis of the coalescence of Ferguson plots of standards and unknown at Y_o, thermostated tube (Fig. 5) or partitioned horizontal slab [22] apparatus is required, and conditions need to be controlled as specified for the construction of Ferguson plots in PAGE (Section 4.2.4). Once that determination has been made, the characterization of a protein subunit by MW, based on the comparison of its migration distance with that of standards, is most accurately carried out in conventional fashion, using a single vertical slab gel (Section 4.3), with the protein of interest and the standards applied in various channels. To ensure that the voltage drop in the various channels is the same, equal sample volumes and sample ionic strength are required in the various channels. For this second step, neither temperature control to give reproducible R_fs, nor reproducible pore size are needed, since the variation of migration distances of unknowns, standards and buffer ions (front) with temperature are proportional, and since the ratio of migration distances between unknowns and standards is independent of gel concentration. Temperature control at that stage must only provide sufficient Joule heat dissipation to allow for conveniently rapid electrophoresis without band distortion. However, the choice of apparatus is dictated by the size of the sample volumes. Vertical apparatus is advantageous if sample volumes do not exceed a few hundred μL; when sample volumes are less than 100 μL per channel, horizontal slab apparatus is applicable. If sample volumes exceed a few hundred μL per channel, gel tube apparatus is required.

15.5   Gel

The gel concentration in SDS-PAGE is selected to spread zones within the R_f range of 0.25 to 0.85 in order to minimize measurement error. Thus, the choice of gel concentration depends on the selection of the trailing ion mobility in the resolving gel (see Section 15.6). In practice, the gel concentration is fixed at any convenient value between 8 and 12 %T. Crosslinking at 2 %C_{Bis} is to be preferred to 5 %C_{Bis} when one wants to spread R_f values within a narrow range of molecular weights. Then, the desired R_f range is achieved by appropriate choice of the trailing ion mobility, as shown in the following section. For analysis of a wide range of molecular weights, a pore gradient gel is advantageous. But it is imperative that the range of gel concentrations used be compatible with linear Ferguson plots intersecting at Y_o, to be able to correlate migration distance with MW. Thus, at high gel concentrations (e.g. 30 %T) moving boundary (front) retardation, micellar SDS retardation and the difficulty in controlling the polymerization rate and, therefore, in making a homogeneous gel, may perturb the system.

15.6   Buffer System

SDS-PAGE in a continuous buffer has the same advantages with regard to simplicity, and limitations with regard to the requirement for concentrated samples, as discussed for PAGE in general (Chapter 3). SDS-PAGE in discontinuous buffer systems proceeds in any buffer system with operative pH of 7 or above, if one selects the appropriate mobility range between leading and trailing ions. (At least, this appears to be the case for proteins at 25 °C, with pIs of about 5. With proteins of higher pI, the lowest pH selected for SDS-PAGE should be at least 2 pH units above the pI). To stack SDS-proteins, a trailing ion net mobility of 0.300 at 25 °C is tolerable; 0.200 is selected in practice to provide a safety margin [78]. The choice of the trailing ion mobility in the resolving gel depends on the choice of gel concentration (see Section 15.5). In practice, the trailing ion mobility in the resolving gel is varied according to column 2, Fig.4 and Appendix 1 [or column 1 of the Subsystems Table on p.3 of the Extensive MBE Buffer Systems Output (Fig.5 of Appendix 2)] until, at a conveniently high gel concentration, the proteins of interest and standards occupy an R_f range of 0.25 to 0.85.

Since a high value of the trailing ion mobility in the stacking gel is tolerated, a stacking gel in SDS-PAGE can equally be prepared according to the specifications in the Extensive Buffer Systems Output for a GAMMA phase as for a BETA phase. This results in a systems designation with 2 Roman numerals, indicative of the GAMMA phases used as stacking and resolving gels (e.g. 4217.II.IV. in [78]). The properties of the resulting boundaries are indicated on page 2 of the Extensive Output in columns 5 and 7 (Fig.3 of Appendix 2).

In general, buffer systems operative at 0 °C are to be preferred in SDS-PAGE over those at 25 °C, although that sounds absurd at first for two reasons. First, low temperature is an anti-denaturation device in protein chemistry. SDS-proteins, of course, are fully denatured, so the application of that device seems senseless. Secondly, SDS precipitates at 0 °C, making it necessary to fudge in practice by operating at about 5 °C in SDS-PAGE or to replace SDS with lithium dodecylsulfate. The reason for our preference of "cold" systems is that polymerization is more easily controlled, and more highly reproducible, at 0 °C than at 25 °C [78],[74],[17] (see Chapter 5).

15.7   Procedure of Polymerization and Electrophoresis

Polymerization in SDS-PAGE is identical to that in PAGE since it is unnecessary and unwarranted to add SDS to the polymerization mixture as is commonly done. It is unnecessary because the mobilities both of SDS monomer and, at moderate gel concentrations, of micellar SDS are far in excess of those of proteins, so that in the field, SDS added to the upper buffer immediately overtakes the proteins.

In electrophoresis, the sole concern of SDS-PAGE relates to the SDS content of the upper buffer and of the sample. The concentration of SDS in the upper buffer has to be sufficient to saturate the migrating proteins during electrophoresis with SDS. For standard proteins, a level of 0.03% appears sufficient [78]; that value may vary with other proteins and needs to be established by systematic variation of SDS concentrations in the upper buffer. Once the proteins are saturated with the detergent, R_f values become independent of SDS levels. Thus the concentration needs to be determined, at which R_fs become constant. Excessive SDS levels are to be avoided because micellar SDS, always present since the concentrations of the detergent in moving boundaries exceed the CMC of about 8.2 mM [5], stacks in the stacking and the resolving gels with the typical trailing ion mobilities of 0.1 or more. Stacked micellar SDS *extends the stack* in space progressively, pushing the trailing edge of the stack backward into the gel, as more and more SDS migrates from the upper buffer into the gel. Thus, the "extended stack" [69] can cover much of the resolving gel, with protein accumulating as a "single band" at its trailing edge. This mechanism obviously shortens the migration path available for separation intolerably, and requires that SDS loads and levels be minimized in SDS-PAGE.

For the same reason of preventing "extended stacks", the SDS content of the sample should also be as low as necessary for saturation of the proteins with detergent. At loads of more than 40 μL of 1% SDS per cm^2 of gel, leading and trailing edges of the moving boundary separate due to the interposition of micellar SDS between leading and trailing ions. In the stacking gel, pyronin Y-SDS and bromphenol blue, respectively, mark those edges. In the resolving gel, R_f in that situation has to be measured with reference to the leading, not the trailing edge of the stack. That leading edge of the stack cannot be marked by the same *tracking dye* over a wide range of gel concentrations, in contradistinction to PAGE in the absence of detergents [78]. In PAGE in general, dyes with mobilities more than the trailing ion mobility in the resolving gel are "tracking" the moving boundary at low gel concentrations and are unstacked only when at very high gel concentrations their mobilities are sufficiently reduced to drop below the trailing ion mobility. Since the trailing ion mobility is a larger value in SDS-PAGE than in PAGE, this type of unstacking of the dye occurs at relatively lower gel concentrations which are well within the common %T values used in SDS-PAGE. Higher mobilities than those of dyes like bromphenolblue are characteristic for cationic dyes after derivatization with SDS (e.g. Pyronin Y-SDS, toluidine blue-SDS). These will be effective tracking dyes at higher gel concentrations than bromphenolblue, but will also unstack as %T is increased. Since these dyes, after reaction with micellar SDS, are macromolecular, their degree of retardation through molecular sieving, once it occurs, is very steep, so that at high gel concentrations unstacked Pyronin Y-SDS may migrate behind stacked bromphenol blue [78]. A further problem in application to "extended stacks", formed by use of excessive amounts of SDS as shown above, is that even stacked dyes may be displaced from the real "front" at the leading edge of the stack by micellar SDS, in view of the alignment of constituents in order of mobility within the stack, to the trailing edge which recedes into the gel as a function of electrophoresis time. Thus, the measurement of R_f in SDS-PAGE is hazardous if one relies on dyes to mark the "front". By inspection, a stacked dye can be clearly distinguished from an unstacked one only in the absence of a substantially extended stack, i.e. when the system contains relatively little SDS. When the stack is extended, the stacked dye may be displaced by micellar SDS from the leading edge – the true front – to the trailing edge of the stack. Thus, the only safe way to locate the true front at the leading edge of an extended stack is by precipitating SDS, either by 0.1 M $BaCl_2$ or by cooling to 0 °C, or by staining it with a cationic dye such as toluidine blue. It appears quite likely in view of these problems in the recognition of the "front" that reported "non-linear" Ferguson plots in SDS-PAGE in discontinuous buffer systems have a good chance to be inaccurate.

15.8   Staining of SDS-Proteins

SDS-PAGE poses problems in protein staining not encountered in PAGE due to reaction of the dye with SDS, due to the competition between SDS and dye for the same sites on the protein, and due to the slowness of diffusing macromolecular SDS-micelles from the gel. These problems require removal of the bulk of SDS, in monomeric form, from the gel prior to, or concomitant with, staining for protein. Commonly, this is done by use of dye solutions in methanolic acetic acid solutions. Their alcohol content increases the CMC of SDS, in addition to the dilution of SDS by the stain solution. Unfortunately, however, lengthy destaining of the gel is required under these conditions to obtain a clear background, and – worse – the efficacy of the methanolic acetic acid solutions of the dye as a protein fixative is insufficient for many proteins, in particular proteins that are hydrophobic, leading to loss of protein bands from the gel during staining.

The solution to these problems appears to lie in the removal of monomeric SDS from the gel *prior* to staining, using a reliable protein fixative like TCA throughout the procedure [55]. Using 12.5% TCA, 45% methanol, monomeric TCA is obtained by increase in the CMC as noted above, while at the same time fixation is assured for probably all types of protein. To accelerate the diffusion of SDS from the gel, the gel may be placed into a perforated drum or a diffusion destainer (Fig.4 of Appendix 4) while a large volume of fixative is magnetically stirred. After a few hours of stirring, pyronin Y-SDS disappears from the gel. The gel may then be stained within 0.5 to 1.0 h by one of the procedures described for PAGE which give no stained background (Section 7.8.1). With a total staining time of 5 h, this procedure just allows one to evaluate SDS-gels on the day of electrophoresis.

15.9   Computation of MW

We have already pointed out in Section 15.1 that a Ferguson plot in SDS-PAGE is required initially to determine whether or not all proteins under analysis, standards and unknowns alike, are derivatized to the same degree by SDS and therefore share a common Y_0 value. We have also shown that if that condition is fulfilled, R_f values at anyone gel concentration can be used to determine MW. We now make the assumption that this condition has been fulfilled in the particular protein system under investigation.

The common procedure of characterization of the SDS-subunits of proteins by SDS-PAGE in terms of MW consists of the electrophoresis of unknowns and MW standards on a single gel slab. The choice of apparatus depends on sample volumes (see Section 15.4). Commonly, the MW of an unknown is obtained by interpolation between the migration distances of MW standards. However, this is correct only when the interpolation is carried out at a logarithmic scale, in view of the logarithmic relationship between migration distance and MW. Improved between-experiment-variance is obtained when the computation of MW is based on R_fs measured on the stained zones (Section 10.3.4), and log MW *vs* R_f or log R_f *vs* MW are plotted for MW standards and unknown to obtain the MW of the latter. Both types of plot are equivalent [113]. Both are sigmoidal. Thus, they can be either evaluated by non-linear curve fitting methods (program SIGMOID), or by linear regression analysis under the assumption of linearity for the central segment of the sigmoidal curves (program GIANTRUN). In both non-linear and linear evaluation of log MW *vs* R_f, computation is required if MW values with their statistical confidence limits, obtained by appropriate consideration of the uncertainty in the line as well as the uncertainty in the data points and by use of appropriate weighting functions, are desired. We will only consider the linear curve fitting approach explicitly in this section. Programs for non-linear curve fitting are available upon request from D. Rodbard, NIH, Bethesda MD 20205.

On line 3 of the input file DATA03 (Fig.13), the term RF is entered to specify the type of function on which MW computation by program GIANTRUN is to be carried out, i.e. log MW *vs* R_f and log R_f *vs* MW.

The standard proteins, each defined by a protein number in data file RADII4 (Fig.30), and their R_f values are entered into data file DATA03 (lines 4 ff., Fig.13). If the standard proteins are not listed in RADII4, they may also be given an available protein number and be entered into file DATA03 together with their molecular radius (from program RADIUS, Appendix 6) and weight.

Program GIANTRUN then computes the unknown MW with its 95% confidence limits and provides an output identical to that described in Fig.14 and Section 10.5.3 except for the plots of the different standard curves on pages 2 and 4 of the output of GIANTRUN (Fig.15).

16 Multicomponent Analysis

When the simultaneous analysis in terms of a gel pattern of many proteins is required, as in the construction of protein maps descriptive of tissue extracts, neither an optimization of the pH of gel electrophoresis, nor of pore size, is possible, in view of the diversity in size and net charge of a multiplicity of proteins. Thus, for multicomponent systems, a charge fractionation applicable to the entire pH range is required, as well as a size fractionation in either a medium pore size, moderately effective on the entire range of molecular sizes under investigation, or in a pore gradient. When one is interested only in the protein subunits (gene products), it is relatively easy to produce such an "average pore size" or pore gradient, since they occur in a rather limited size range, between about 10000 and 100000. For an analysis of (post-translational) proteins in biological systems the range is much wider and the choice of gel concentrations more problematic.

Multicomponent systems may be resolved efficiently by EF at high voltage in a single dimension of a charge fractionation yielding hundreds of protein components [23]. But such analysis necessarily fails to exploit the size differences among proteins for resolution. Single dimensional size fractionations seem much less efficient (presumably in view of the absence of the regulated high protein concentrations operative in EF). It has been estimated that a gel of conventional dimensions on purely geometric grounds is incapable of resolving more than about 50 equiconcentrated and equidistant zones [27]. Therefore, size fractionations alone, or size fractionations coupled to charge fractionations of multicomponent systems must be carried out either in two (or more) stages [27],[29] or in two dimensions [173]. At present, only the 2-dimensional (2D-[EF-SDS-PAGE]) approach [173] has gained wide popularity and will be discussed here. However, both the 2-stage and the 2-dimensional approach to multicomponent resolution rest on the principle that they are capable of revealing the *product* of the numbers of zones obtained in each of the separate stages or dimensions. Thus, assuming 50 equidistant and equiconcentrated components in each of the 2 dimensions, $50 \times 50 = 2500$ spots would fit onto a 10×10 cm square area. The order in which charge or size fractionation are carried out is irrelevant in principle, although practical reasons for preferring one order of the other exist (see below).

16.1 Prerequisites of Multicomponent Resolution in Two Dimensions with Regard to the Fractionation Principle and to Geometry

Two-dimensional analysis has to exploit different principles of fractionation in each of the 2 dimensions, such as charge *vs* size fractionation, because to the degree that each dimension of analysis responds to a single such principle, a spot distribution along the diagonal would result, which, being only 1.4 times longer than each of the original dimensions of fractionation, could not allocate significantly more components than a 1-dimensional analysis. By contrast, an *even* distribution of bands in each of the 2 dimensions of fractionation is capable of fully utilizing the area of the square or rectangle in a 2-dimensional analysis. The resolution of multicomponent systems therefore depends on finding conditions of both charge fractionation and size fractionation in a single dimension which provides an even band distribution across the gel. Optimally, these band distributions should be independent of electrophoresis time, i. e. be derived from steady-state methods in both dimensions.

16.2 Charge Fractionation

The charge fractionation methods available for the first dimension of fractionation are EF and MBE. They need to be the 1st dimension only if, as is commonly the case, the 2nd dimension involves SDS-PAGE, since in the reverse order of application EF or MBE would have to quantitatively separate the proteins from SDS before a charge fractionation would be possible. This would certainly require extremely prolonged fractionation times and probably not be feasible quantitatively at all within practical limits of time and migration path. However, if the 2nd dimensional size fractionation involves PAGE at very high or very low pH (where the charge fractionation element in PAGE can be neglected), the reverse order is equally applicable in principle. In practice, however, where the 2nd dimensional gel is frequently a vertical gel slab, this order of procedural steps would require a solution to those problems that have to date prevented the wide application of vertical gels on slab EF (see Section 12.1).

MBE has not been applied to date to the 1st dimension of 2-dimensional gel fractionations although it may very well lend itself to an even zone distribution whenever the components of a multicomponent system occur in approximately the same concentrations and therefore give rise to zones of similar lengths. Such application of MBE will also depend on very sensitive detection methods, since the steady-state of MBE can practically only be reached in application to multicomponent systems if the protein load is very small (see the introduction to Chapter 13). The contiguous alignment of zones in MBE is likely to be compatible with its use in the 1st dimension, since in the 2nd dimension the contiguous components are separated. MBE has the further advantage compared to EF that the separated proteins are in a soluble form. Potentially it should be useful that the length of sequential buffer zones in MBE can be computed (once their ionic mobilities have been measured) to fit the train of buffer compartments into the desired geometry of the 2-D apparatus; proteins with intermediate net mobilities between the buffers would align in the assigned length of gel (assuming negligible amounts of protein compared to the buffers) or would extend it to a degree that could be corrected by changing the length of the buffer compartments [115].

In present practice of 2-dimensional gel fractionation, EF has nearly always been used as the 1st dimension. This application not only burdens the 2-dimensional analysis with the inherent problems of gel EF discussed in Chapter 12, i.e. the need to reach the steady-state with regard to all component proteins, the problem of protein-protein and protein-carrier ampholyte interactions, and the insolubility of the isoelectric proteins, but also raises additional problems:

16.2.1   Even Zone Distribution

pH Gradient engineering has to achieve an even zone distribution across the pH gradient. This requires a stretching of the pH 5 region across much of the pH gradient, since most proteins are isoelectric around pH 5. Present empirical approaches to achieve such pH gradients have to employ all of the trickery of constituent displacement, of constituent addition and subtraction, of buffer EF, and of pH gradient stabilization discussed in Chapter 12. But these are obviously too laborious for the purpose of providing an even zone distribution efficiently. An efficient solution to this problem would require a computer simulation of pH gradients and their dynamics which is available potentially for electrofocusing in simple buffers (not in SCAMs) but not practical as yet in the absence of ionic mobility data for most buffer constituents [195].

16.2.2   Detergent Problems

Most 2-dimensional EF-SDS-PAGE is carried out in 8 – 9.5 M urea and non-ionic detergent, without considering a) the transport of non-ionic detergent in electrofocusing; b) the physical state of the detergent (monomeric or micellar); c) the effect of urea and other perturbing factors on that physical state; d) the binding of SCAMs to the detergent; e) the physical state of the protein-detergent interaction product at the steady-state of electrofocusing. This state is of importance for 2-dimensional electrophoresis because it defines the solubility and availability for reaction with SDS of the isoelectric zone.

a) Non-ionic detergent micelles may associate with charged detergents, carrier constituents or proteins and migrate electrophoretically with them. An example is the association of Lubrol PX with either deoxycholate or N-laurylsarcosinate (Fig.3 of [4]). Necessarily, such migration leaves some gel region deficient in the detergent concentration required for protein solubility when the protein is water-insoluble, and must lead to protein precipitation with consequent problems of local high voltage and artifactual banding.

b) Low CMC detergents like Triton X-100 (alias NP-40) or Lubrol PX (Table 1 of [3]) exist at the concentrations used in EF (about 0.1%) in forms of micelles with micellar weights of about 100000. These associate with the hydrophobic protein to give complexes (possibly of several homologous or heterologous proteins with one or several detergent micelles) which are very large. They are consequently retarded in the EF gel (unless its composition is tailored to being non-restrictive to such complexes) and therefore cannot attain their isoelectric positions on the pH gradient within the allotted time.

c) Urea, ethylene glycol increase, and sucrose decreases the CMC of Triton X-100 (Fig.8 of [5]). Thus, detergent micelles may not be available in cases where they are needed for solubilization. It appears safer, therefore, to apply urea in conjunction with those high CMC detergents like CHAPS or β-octylglucoside which appear to solubilize membrane proteins below their CMC [3].

d) Basic SCAMs being hydrophobic, they may bind detergents and hydrophobic proteins selectively. An example is the binding of Triton X-100 in the alkaline region of pH gradients (Fig. 4 of [139]).

e) As the protein approaches its pI position on the pH gradient, and particularly, as the EF time is increased beyond the minimum required to reach a constant (i.e. near-isoelectric) pH, proteins tend to precipitate irreversibly.

A discussion of the particular problems of multicomponent analysis, in both dimensions considered separately and jointly, is found in [174].

16.3   Size Fractionation

SDS-PAGE is the commonly used 2nd dimensional size fractionation tool. It has in that application the fundamental advantage that the reaction of SDS with the insoluble isoelectrically precipitated zones of the 1st dimensional EF tends to solubilize them. Furthermore, the elimination of net charge differences by reaction with SDS eliminates trivial charge heterogeneity and improves thereby the reproducibility of migration distances. For probably the same reason, band width of SDS-derivatives is less than that of the native proteins [30]. It has, however, also a number of inherent problems: a) Differences in native aggregation or association state and conformation as well as biological activities are lost. This decreases the efficiency of resolution through loss of some of the criteria by which molecules are differentiated. It also abolishes the possibility to obtain a biologically relevant pattern of the *in vivo* post-translational states of molecules. b) There is no safe way to saturate the proteins under the EF zones with SDS under the conventional conditions of SDS-PAGE.

SDS-PAGE in the second dimension shares with the application of PAGE at an extreme of pH, applied for the same purpose, some further problems: c) Stacking in the 2nd dimension of the solubilized isoelectric zones of the 1st dimension needs to be verified, to assure simultaneous arrival of proteins at the starting zone of the resolving gel in spite of different solubilization rates. d) Stacking gels must not be polymerized around the EF gel, to avoid substantial retention in the EF gel of protein migrating into the 2nd dimension [175].

16.3.1   Loss of Protein Functions

The denaturing and dissociating effect of SDS binding to proteins deprives 2-dimensional electrophoresis of one of its main functions in biology and medicine: The possibility to construct macromolecular maps of native species. An alternative to the use of SDS is to select for the 2nd dimension a pH sufficiently high or low to solubilize the isoelectric proteins of the 1st dimension, and allow them to migrate unidirectionally and to stack while at least partially maintaining native functions. To include hydrophobic proteins on such a 2-D map of native proteins, at least partially non-denaturing non-ionic or amphoteric detergents should be included in either one of the 2 dimensions [1],[3].

16.3.2   Saturation with SDS in the Cold

The conditions under which proteins acquire equal charge densities through reaction with SDS involve heating to 100 °C for variable times, and quantitative reduction of disulfide bonds. It is patently impossible to subject a 1st dimensional gel to those conditions, since proteins must diffuse out of the gel under conditions of boiling with SDS, and since the oxidative 1st dimensional gel has to be fully reduced before the proteins can start to react; this requires high concentrations of reducing agent, and concentrations which vary with gel concentration and pH (see Section 5.1). In the practice of the 2-D method, it has been found that at least the uncoiling and chain separation brought about by heat can in many cases be achieved by solvents such as 9 M urea-2% Triton X-100 [174]. Similar effectiveness, without the problems associated with low CMC and high aggregation number of the detergent, may be expected from mixtures of urea with CHAPS or β-octylglucoside. However, a procedure for cleavage of disulfide bonds in these solvents remains to be worked out before one could safely conclude that saturation of proteins with SDS is feasible at ambient temperatures.

16.3.3   Stacking of Resolubilized Isoelectric Proteins

Solubilization rates for the 1st dimensional isoelectric protein precipitates must differ. Thus, it is necessary in the 2nd dimension, not only to ascertain that the range of net mobilities between leading and trailing ions of a stack is wide enough to accomodate all 1st dimensional proteins, by running them into a stacking gel and staining, but also to experimentally determine a stacking gel migration path length sufficient for providing a simultaneous entrance of all proteins into the resolving gel.

16.3.4   Gelling-in of the First Dimensional Gel

When 1st dimensional gels are polymerized into polyacrylamide stacking gels, proteins become unable to migrate into the 2nd dimensional gel while they do so freely when the stacking gel of the 2nd dimension is made of agarose [175]. Although the mechanism of this phenomenon remains unknown, one may speculate that it is due to the participation of the residual monomers in the gel in the growth of polymer chains in the polymerization mixture surrounding it. Such participation could produce molecular sieving effects at the surface of the gel strong enough to prevent the migration of proteins from the 1-D gel.

16.4   Apparatus

The common apparatus of 2-dimensional gel electrophoresis comprises a tube apparatus for EF in the first dimension and a vertical slab apparatus for the second. The tube gel is physically transferred onto the surface of the slab, with or without pre-reaction with SDS, and is subjected to SDS-PAGE. To date, a single apparatus for electrophoresis in both dimensions still needs to be developed. Two concepts should be mentioned which may aid in its future development, *viz.* 1) the fundamental identity of EF and MBE [20],[115]; and 2) the possibility to replace sealed upper and lower buffer reservoirs by upper and lower buffers differing in density, with interposition of an immiscible solvent of intermediate density (Matthews, G., unpublished data).

16.4.1 Utilization of the Fundamental Identity between Stacking and Electrofocusing

Stacking gels subjected to electrophoresis using an upper buffer containing a mixture of trailing constituents with a range of net mobilities, give rise at the steady state to an alignment of these constituents behind the leading (gel) buffer species. A pH gradient crosses the aligned array of species. By choice of the counterion concentration, the constituents aligned in a stack, and thus the pH gradient, can be decelerated to any desired degree. When the counterion concentration is zero, and protons and hydroxyl ions are the sole counterions in the system, the pH gradient is stationary. When the counterion concentration is elevated, the pH gradient is mobilized. Thus, one could set up in the same gel an arrested pH gradient, turn the gel slab 90° and continue electrophoresis with the trailing ion of the lowest mobility as upper buffer. This would at a new steady state stack all species in the 2nd dimension, providing that the gel contains a large excess of the leading ion. Such interconversion between stacking and electrofocusing gels has been experimentally demonstrated (although at that time necessarily in ignorant and clumsy fashion) [176] long before the theoretical relation between stacking and electrofocusing was understood [20],[115].

16.4.2 Elimination of Electrolyte Chambers

PAGE can be conducted on a square gel slab in a simple beaker containing lower buffer in 20% sucrose, an intermediate layer of dibutylphthalate (density 1.05) and aqueous upper buffer. This arrangement allows one to use the slab gel first in EF, to replace the electrolyte solutions and rotate the slab by 90 °C, and to proceed with electrophoresis in the second dimension without any physical gel transfer.

The size of 2-D gels should be as large as practicable. It is obvious that resolution of multicomponent systems is proportional to gel surface area.

16.5 Protein Detection

Protein detection on macromolecular maps of several hundred or thousand components depends on detection sensitivity, and on obtaining uniformly small spots.

16.5.1   Detection Sensitivity

Autoradiograms of radioactively labelled multicomponent systems do better in this regard than protein staining patterns, by providing at least 5 times more detectable components [173]. Similarly, the silver staining method [177] is at least by one order of magnitude more sensitive than ordinary protein staining methods in gel electrophoresis; a particularly useful form of the silver stain in application to 2-dimensional gels appears a modification yielding multicolored patterns [178]. However, these advantages in sensitivity of autoradiography or silver stain simply relate to the low concentrations of many components in the sample. If the sample load is increased, less sensitive detection methods are needed to provide 2-dimensional patterns with spot sizes small enough to fit as many spots as possible onto the available gel area.

16.5.2   Overshadowing

An important problem in 2-dimensional spot detection is the overshadowing of less abundant components by the more highly concentrated ones. Therefore, non-linear color values or non-linear photographic response are valuable, since they can provide approximately equally small spots for more concentrated and less concentrated components of the pattern. In this regard, autoradiography may excel over staining methods, particularly since it allows one to analyze the same gel pattern at different times of exposure.

16.6   Detection of a Single Pattern Change Within a Multicomponent Pattern

The single pattern changes within complex 2-dimensional spot patterns have been evaluated initially by visual comparison and necessarily subjectively [173]. More recently, the densitometric and automated pattern analysis of macromolecular maps has allowed for an objective spot identification (e.g. [179]-[183]). However, due to gel irreproducibility, this is not possible in a rigid coordinate system but rather requires a shifting coordinate system centered on the dominant spots of the pattern. Furthermore, a single spot alteration in a complex pattern to be significant still requires identification by several independent biochemical or genetic techniques [174]. Thus, while the physical tools for spot identification on 2-dimensional macromolecular maps have been highly developed, the reproducibility of gels, apparatus and procedure of 2-D analysis need to be developed further, before the valid grand dream of clinical and biological macromolecular mapping can be realized.

Acknowledgements

While it is correct, as stated in the Preface, that the biochemical practitioner should not be concerned with fractionation theory but rather should, in order to reap the benefits of theoretical and statistical "quantitative" methods, take advantage of available computer programs and computer output, it must also be remembered that without the work of theoreticians and programmers, an applied quantitative gel electrophoresis would not be possible. The fundamental theoretical work and the computer programs on which this entire book and the approach to quantitative gel electrophoresis we have been trying to communicate rest, is due to Dr. T. M. Jovin (Max Planck Institut fuer Physikalische Chemie, Goettingen) with respect to multiphasic buffer systems, and to Dr. D. Rodbard (NIH, Bethesda MD) with regard to pore theory and the related statistics. My present understanding of detergent-PAGE and of the physical-chemical aspects of electrofocusing derives largely from the original insights of Dr. L. M. Hjelmeland (NIH, Bethesda MD). The numerous contributions made by my other friends and collaborators in the work which is summarized by this book are reflected by the bibliography. Their first authorship on most papers testifies to the fact that they, and not I, have made most of the findings. Also, since ideas arise from human interactions, the concepts advanced in this book are indistinguishably theirs and mine.

References

[1] Hjelmeland, L. M., and Chrambach, A. (1981),"Electrophoresis and Electrofocusing in Detergent Containing Media: A Discussion of Basic Concepts" *Electrophoresis* **2**, 1-11.

[2] Newby, A. C., Chrambach, A., and Bailyes, E. M. (1982), "The Choice of Detergents for Molecular Characterization and Purification of Native Intrinsic Membrane Proteins" in: *Techniques in Lipid and Membrane Biochemistry – Part I* (Hesketh, T. R., Kornberg, H. L., Metcalfe, J. C., Northcote, D. H., Pogson, C. I., and Tipton, K. F., eds.) Vol. B4/I, pp. 1-22. Elsevier/North Holland Biomedical Press.

[3] Hjelmeland, L. M., and Chrambach, A. (1983), "Solubilization of Functional Membrane Bound Receptors", in: *Receptor Biochemistry and Methodology* (Venter, J. C., and Harrison, L., eds.) Vol. 1, A, pp. 35-46. R. Liss Inc., New York.

[4] Newby, A. C., and Chrambach, A. (1978), "Disaggregation with Enhanced Net Charge of Enzymatically Active Adenylate Cyclase in Mixtures of Ionic and Non-ionic Detergents", *Biochem. J.* **177**, 623-630.

[5] Hjelmeland, L. M., Nebert, D. W., and Chrambach, A. (1978), "Electrophoresis and Electrofocusing of Native Membrane Proteins", in *Electrophoresis '78*, pp. 29-56, (Catsimpoolas, N., ed.). Elsevier, North Holland Publ. Co., Amsterdam-New York NY.

[6] Hjelmeland, L., Nebert, D., and Osborne, J. (1983), "Sulfobetaine Derivatives of Bile Acids: Nondenaturing Surfactants for Membrane Biochemistry", *Anal. Biochem.* **130**, 72-82.

[7] Chen, B., Rodbard, D., and Chrambach, A. (1978), "Polyacrylamide Gel Electrophoresis with Optical Scanning, Using Multiphasic Buffer Systems The Stack", *Anal. Biochem.* **89**, 596-608.

[8] Hjelmeland, L. M., and Chrambach, A. (1982), "The Impact of L. G. Longsworth on the Theory of Electrophoresis", *Electrophoresis* **3**, 9-17.

[9] Ornstein, L. (1964), "Disc Electrophoresis -I Background and Theory", *Ann. N. Y. Acad. Sci.* **121**, 321-349.

[10] Jovin, T. M. (1973), "Multiphasic Zone Electrophoresis.I,II,III", *Biochemistry* **12**, 871, 879, 890.

[11] Everaerts, F. M., Beckers, J. L., and Verheggen, T. P. E. M. (1976), "Isotachophoresis: Theory, Instrumentation and Applications", Elsevier, Amsterdam.

[12] Schafer-Nielsen, C., and Svendsen, P. J. (1981), "A Unifying Model for the Ionic Composition of Steady-State Electrophoresis Systems", *Anal. Biochem.* **114**, 244-262.

[13] Jovin, T. M., (1973), "Multiphasic Zone Electrophoresis.IV Design and Analysis of Discontinuous Buffer System with a Digital Computer", *Ann. N. Y. Acad. Sci.* **209**, 477-496.

[14] Jovin, T. M., Dante, M. L., and Chrambach, A., *Multiphasic Buffer Systems Output*, National Technical Information Service, Springfield, Virginia 22151, 1970, PB 196085 to 196091, 259309 to 259312, 203016.

[15] Chrambach, A., and Jovin, T. M. (1982), "Selected Buffer Systems for Moving Boundary Electrophoresis at Various pH Values and at Both Polarities, Presented in a Simplified Format", *Electrophoresis* **4**, 190-204.

[16] Rodbard, D., and Chrambach, A. (1971), "Estimation of Molecular Radius, Free Mobility, and Valence Using Polyacrylamide Gel Electrophoresis", *Anal. Biochem.* **40**, 95-134.

[17] Chrambach, A., Jovin, T. M., Svendsen, P. J., and Rodbard, D. (1976), "Analytical and Preparative Polyacrylamide Gel Electrophoresis: An Objectively Defined Fractionation Route, Apparatus and Procedures", in: *Methods of Protein Separation*, (Catsimpoolas, N., ed.), Vol.2, pp. 27-144. Plenum Press, New York NY.

[18] Chrambach, A., and Rodbard, D. (1980), "'Quantitative' and Preparative Polyacrylamide Gel Electrophoresis", in: *Gel Electrophoresis of Proteins: A Practical Approach*, pp. 93-144 (Hames, B. D., and Rickwood D., eds.). IRL Press Ltd., London and Washington DC.

[19] Routs, R. J. (1971), "Electrolyte Systems in Isotachophoresis and Their Application to Some Protein Separations", *Doctoral Thesis*. Technische Hogeschool Te Eindhoven, pp. 1-137. Solna Skriv-& Stenograftjaenst AB, Solna, Sweden.

[20] Hjelmeland, L. M., and Chrambach, A. (1983), "Formation of Natural pH Gradients in Sequential Moving Boundary Systems with Solvent Counterions (I): Theory", *Electrophoresis* **4**, 20-26.

[21] Buzas, Zs., and Chrambach, A. (1982), "Un-Supercoiled Agarose With a Degree of Molecular Sieving Similar To That of Crosslinked Polyacrylamide", *Electrophoresis* **3**, 130-134.

[22] Serwer, P. (1983), "Improved Procedures for Controlling the Voltage Gradient and Temperature During Electrophoresis in Submerged Horizontal Slab Gels", *Electrophoresis* **4**, 227-231.

[23] Allen, R. C. (1980), "Rapid Isoelectric Focusing and Detection of Nanogram Amounts of Proteins from Body Tissues and Fluids". *Electrophoresis* **1**, 32-36.

[24] Davis, B. J. (1964), "Disc Electrophoresis-II Method and Application to Human Serum Proteins", *Ann. N. Y. Acad. Sci.* **121**, 121.

[25] Neuhoff, V. (1970), "Die Disk-Elektrophorese als Analytische Mikromethode zur Charakterisierung von Makromolekülen", *Mitt. Dtsch. Pharmaz. Ges.* **12**, 289-314.

[26] Chrambach, A., Hearing, E., Lunney, J., and Rodbard, D. (1972), "Experimental Validation of the Predicted Properties of a Multiphasic Buffer System Applied to Polyacrylamide Gel Electrophoresis", *Separation Sci.* **7**, 725-745.

[27] Kapadia, G., Chrambach, A., and Rodbard, D. (1974), "Approaches to Macromolecular Mapping by Polyacrylamide Gel Electrophoresis", in: *Electrophoresis and Electrofocusing on Polyacrylamide Gel*, pp. 115-144, (Allen, R. C., and Maurer, H. R., eds). W. de Gruyter, Berlin-New York NY.

[28] Muniz, N., Rodbard, D., and Chrambach, A. (1977), "Identification of Macromolecules by Quantitative Polyacrylamide Gel Electrophoresis A Comparison of 3 Approaches", *Anal. Biochem.* **83**, 724-738.

[29] Altland, K., and Hackler, R. (1981), "Horizontal Gradient Polyacrylamide Gel Electrophoresis", *Electrophoresis* **2**, 49-54.

[30] Chen, B., and Chrambach, A. (1980), "The Effect of SDS on Protein Zone Dispersion in Polyacrylamide Gel Electrophoresis", *Anal. Biochem.* **102**, 409-418.

[31] Maxam, A. M., and Gilbert, W. (1977), "A New Method for Sequencing DNA, *Proc. Natl. Acad. Sci. USA* **74**, 560-564.

[32] Rapaport, R. N., Jackiw, A., and Brown, R. K. (1980), "pH Gradient Flattening in Isoelectric Focusing in Long Polyacrylamide Gels", *Electrophoresis* **1**, 122-126

[33] Radola, B. (1980), "Ultrathin-Layer Isoelectric Focusing in 50-100 μm Polyacrylamide Gels on Silanized Glass Plates or Polyester Films", *Electrophoresis* **1**, 43-56.

[34] Frey, M.D. and Radola, B.J. (1982), "High Resolution Preparative Isoelectric Focusing in Layers of Granulated Gels", *Electrophoresis* **3**, 216-226.

[35] Cuono, C., and Csapo, G. (1981), "Gel Electrofocusing in a Natural pH Gradient of pH 3-10 Generated by a 47-Component Buffer Mixture", *Electrophoresis* **2**, 65-75.

[36] Chrambach, A., An der Lan, B., Mohrmann, H., and Felgenhauer, K. (1981), "Toward an Improved Immunoglobulin Analysis by Gel Electrophoresis and Electrofocusing", *Electrophoresis* **2**, 279-286.

[37] Nguyen, N. Y., McCormick, A. G., and Chrambach, A. (1978), "Anodic Drift of pH Gradients in Electrofocusing on Polyacrylamide Gel", *Anal. Biochem.* **88**, 186-195.

[38] Righetti, P.G., Gianazza, E., and Bianchi Bosisio, A. (1979), "Biochemical and Clinical Applications of Isoelectric Focusing", in: *Recent Developments in Chromatography and Electrophoresis*, pp. 1-36 (Frigerio, A., and Renoz, L., eds.). Elsevier, Amsterdam, The Netherlands,

[39] Serwer, P. (1980), "A Technique for Electrophoresis in Multiple-Concentration Agarose Gels", *Anal. Biochem.* **101**, 154-159.

[40] Serwer, P. (1983), "Agarose Gels: Properties and Use for Electrophoresis", *Electrophoresis* **4**, 375-382.

[41] Newby, A. C., Matthews, G., and Chrambach, A. (1978), "A Simplified Polyacrylamide Gel Electrophoresis Apparatus for Simultaneous Application of Multiple Buffer Systems or Detergent Combinations", *Anal. Biochem.* **91**, 473-480.

[42] Chrambach, A. (1980), "Electrophoresis and Electrofocusing on Polyacrylamide Gel in the Study of Native Macromolecules", *J. Mol. Cell Biochem.* **29**, 23-46.

[43] Rodbard, D., Chrambach, A., and Weiss, G. H. (1974), "Optimization of Resolution in Analytical and Preparative Polyacrylamide Gel Electrophoresis", in: *Electrophoresis and Electrofocusing on Polyacrylamide Gel*, pp. 62-105 (Allen, R. C., and Maurer, H. R., eds.). W. de Gruyter, Berlin-New York NY.

[44] Rodbard, D., Kapadia, G., and Chrambach, A. (1971), "Pore Gradient Electrophoresis", *Anal. Biochem.* **40**, 135-157.

[45] Felgenhauer, K. (1979), "Characterization of Native Multicomponent Protein Mixtures by One- and Two-Dimensional Gradient Electrophoresis", *J. Chromatogr.* **173**, 299-311.

[46] Margolis, J., and Kenrick, K. G. (1968), "Polyacrylamide Gel Electrophoresis in a Continuous Molecular Sieve Gradient", *Anal. Biochem.* **25**, 347-362.

[47] Coleman, P. F., and Gabriel, O. (1982), "Lipopolysaccharides", in: *Electrophoresis, a Survey of Techniques and Applications. Part B – Applications,* pp. 281-286. (Deyl, Z., Chrambach, A., Everaerts, F. M., and Prusik, Z., eds.). Elsevier, Amsterdam-Oxford-New York.

[48] Chrambach, A., A., Pickett, J., Schlam, M. L., Kapadia, G., and Holtzman, N. A. (1972), "A Partitioned Slab Apparatus for One-Dimensional Polyacrylamide Gel Electrophoresis in Multiphasic Buffer Systems Under a Wide Range of Conditions", *Separation Sci* **7**, 773-783.

[49] Chidakel, B. E., Ellwein, L. E., and Chrambach, A. (1978), "Controlled Deaeration of Polymerization Mixtures in Polyacrylamide Gel Electrophoresis", *Anal. Biochem.* **85**, 316-320.

[50] Kapadia, G., and Chrambach, A. (1972), "Recovery of Protein in Preparative Polyacrylamide Gel Electrophoresis", *Anal. Biochem.* **48**, 90-102. Appendix II: Preparative PAGE Procedure for Apparatus B. Appendix III: Preparative PAGE Procedure for Apparatus C and D.

[51] Oster, G., Bellin, J. S., and Holmstroem, B. (1962), "Photochemistry of Riboflavin", *Experientia* **18**, 249-253.

[52] Ben-Or, S., and Chrambach, A. (1981), "Multiple Forms of Glucocorticosteroid Receptors in the Neural Retina of the Chick Embryo, Revealed by Polyacrylamide Gel Electrophoresis", *Arch. Biochem. Biophys.* **206**, 318-330.

[53] Nguyen, N. Y., and Chrambach, A. (1980), "Analytical and Preparative Gel Isotachophoresis of Macromolecules", in: *Gel Electrophoresis of Proteins: A Practical Approach,* pp. 145-155. (Hames, B. D., and Rickwood D., eds.). IRL Press Ltd., London-Washington DC.

[54] An der Lan, B., Ben-Or, S., Chrambach, A., Allenmark, S., and Jackiw, B. A. (1984), "Gel Electrofocusing in Tubes Containing Sephadex", *Electrophoresis* **5**, 343–348.

[55] An der Lan, B., Sullivan, J. V., and Chrambach, A. (1982), "Effective Fixation and Rapid Staining of Proteins in SDS-PAGE", in: *Electrophoresis '82,* pp. 225-233. (Stathakos, D., ed.). W. de Gruyter, Berlin-New York NY.

[56] An der Lan, B., and Chrambach, A. (1981), "Analytical and Preparative Gel Electrofocusing", in: *Gel Electrophoresis of Proteins: A Practical Approach,* pp. 157-187. (Hames, B. D.. and Rickwood D., eds.). Information Retrieval Ltd. Press, London and Washington DC.

[57] Chrambach, A., Hjelmeland, L., Nguyen, N. Y., and An der Lan, B. (1980), "Gel Electrofocusing With Increased Degrees of Freedom", in: *Electrophoresis '79,* pp. 3-22. (Radola, B. J., ed.). W. de Gruyter, Berlin-New York NY.

[58] Peterson, J. I., H. W. Tipton, and A. Chrambach, (1974), "A Gel Slicer for the Transverse Sectioning of Cylindrical Polyacrylamide Gels", *Anal. Biochem.* **62**, 274-280.

[59] Chrambach, A. (1966), "A Device for the Sectioning of Cylindrical Polyacrylamide Gels", *Anal. Biochem.* **15**, 544-548.

[60] Kohler, P. O., Bridson, W. E., and Chrambach, A. (1971), "Human Growth Hormone Produced in Tissue Culture: Characterization by Polyacrylamide Gel Electrophoresis", *J. Clin. Endocr.* **32**, 70-76.

[61] Chidakel, B. E., Baumann, G., Rodbard, D., and Chrambach, A. (1975), "A Simple Electronic Device for the Measurement of Band Migration Distances in Polyacrylamide Gels", *Anal. Biochem.* **66**, 540-544.

[62] Nochumson, S., and Gibson, S.G. (1983), "Oven Drying of High-Percentage Polyacrylamide Slab Gels on Support Films Using a New Cross-Linking Agent., *Biotechniques* **1**, 18-23.

[63] Chidakel, B. E., Nguyen, N. Y., and Chrambach, A. (1977), "A Device for the Automated Measurement of pH-Gradients in Gel Electrofocusing", *Anal. Biochem.* **77**, 216-225.

[64] Nguyen, N. Y., and Chrambach, A. (1979), "A 3-Step Method for Isolating A Few to Several Hundred Milligrams of Protein", *J. Biochem. Biophys. Methods* **1**, 171-187.

[65] An der Lan, B., Horuk, R., Sullivan, J. V., and Chrambach, A. (1983), "An Improved and Simplified Apparatus for Protein Extraction and Concentration from Gel Slices, Using Moving Boundary Electrophoresis", *Electrophoresis* **4**, 335-337.

[66] Chrambach, A., and Nguyen, N. Y. (1978), "Preparative Electrophoresis, Isotachophoresis and Electrofocusing on Polyacrylamide Gel", in: *Electrokinetic Separation Methods*, pp. 337-367. (Righetti, P. J., van Oss, C. J., and Vanderhoff, J. W., eds.). Elsevier, Amsterdam, The Netherlands.

[67] Kapadia, G., Vaitukaitis, J., and Chrambach, A. (1981), "One-Step Isolation of Human Chorionic Gonadotropin in Milligram Amounts, using Selective Steady-State Stacking on Polyacrylamide Gel", *Preperative Biochem.* **11**, 1-22.

[68] Ellwein, L. B., Huff, R. W., and Chrambach, A. (1977), "A Simple Device For the Continuous Deceleration of Eluent Flow in Preparative Polyacrylamide Gel Electrophoresis", *Anal. Biochem.* **82**, 46-53.

[69] Baumann, G., and Chrambach, A. (1976), "Gram-Preparative Protein Fractionation by Isotachophoresis: Isolation of Human Growth Hormone Isohormones", *Proc. Natl. Acad. Sci. USA* **73**, 732-736.

[70] Rodbard, D., and Chrambach, A. (1970), "Unified Theory of Gel Electrophoresis and Gel Filtration", *Proc. Natl. Acad. Sci. USA* **65**, 970-977.

[71] Allegrini, J., Alhenc-Gelas, F., An der Lan, B., and Chrambach, A. (1984), "Ferguson Plot Analysis at the Nanogram Level", Abstr. 2nd Meet. Electrophoresis Soc. USA, Tucson AZ, Oct. 17-19.

[72] Chrambach, A., and Rodbard, D. (1971), "Polyacrylamide Gel Electrophoresis", *Science* **172**, 440-451.

[73] Rodbard, D., Levitov, C., and Chrambach, A. (1972), "Electrophoresis in Highly Crosslinked Polyacrylamide Gels", *Separation Sci.* **7**, 705-723.

[74] Chrambach, A., and Rodbard, D. (1972), "Polymerization of Polyacrylamide Gels Efficiency and Reproducibility as a Function of Catalyst Concentration", *Separation Sci.* **7**, 663-703.

[75] Chen, B., and Chrambach, A. (1979), "Estimation of Polymerization Efficiency in the Formation of Polyacrylamide Gel, Using Continuous Optical Scanning During Polymerization", *J. Biochem. Biophys. Methods* **1**, 105-116.

[76] Stegemann, H., Francksen H., and Macko, V. (1973), "Potato Proteins: Genetic and Physiological Changes Evaluated by One- and Two-Dimensional PAA-Gel-Techniques", *Z. Naturforschung* **28c**, 722-732.

[77] Schenkein, I., Levy, M., and Weis P. (1968), "Apparatus for Preparative Electrophoresis Through Gels", *Anal. Biochem.* **25**, 387-395.

[78] Wyckoff, M., Rodbard, D., and Chrambach, A. (1977), "Polyacrylamide Gel Electrophoresis in Sodium Dodecylsulfate Containing Buffers, Using Multiphasic Buffer Systems: Properties of the Stack, Valid R_f-Measurement and Optimized Procedure", *Anal. Biochem.* **78**, 459-482.

[79] Hjelmeland, L. M., Allenmark, S., An der Lan, B. Jackiw, B. A., Nguyen, N. Y., and Chrambach, A. (1981), "Electrophoresis in Gels Containing Zwitterionic Groups", *Electrophoresis* **2**, 82-90.

[80] Baumann, G., and Chrambach, A. (1976), "A Highly Crosslinked Transparent Polyacrylamide Gel with Improved Mechanical Stability for Use in Electrofocusing and Isotachophoresis", *Anal. Biochem.* **70**, 32-38.

[81] Schildknecht, C. E. (1973), *Allyl Compounds and Their Polymers*, Wiley-Interscience, New York NY.

[82] Ruechel, R., and Brager, M. D. (1975), "Scanning Electron Microscopic Observation of Polyacrylamide Gels", *Anal. Biochem.* **68**, 415-428.

[83] Gressel, J., and Robards, A. W. (1975), "Polyacrylamide Gel Structure Resolved?", *J. Chromatogr.* **11**, 455-458.

[84] Morris, C. J . O. R., and Morris, P. (1971), "Molecular-Sieve Chromatography and Electrophoresis in Polyacrylamide Gels", *Biochem. J.* **124**, 517-528.

[85] White, M. L., and Dorion, G. H. (1961), "Diffusion in a Crosslinked Acrylamide Polymer Gel", *J. Polymer Sci.* **55** , 731-740.

[86] Peacock, A. C., and Dingman, C. W. (1967), "Resolution of Multiple Ribonucleic Acid Species by Polyacrylamide Gel Electrophoresis", *Biochemistry* **6**, 1818-1827.

[87] Uriel, J., and Berges, J. (1974), "Electrophoresis in Polyacrylamide-Agarose Composite Gels: An Outline of the Method and Its Applications", in: *Electrophoresis and Electrofocusing on Polyacrylamide Gel*, pp. 235-245. (Allen, R. C., and Maurer, H. R., eds.). W. de Gruyter, Berlin-New York NY.

[88] Ferguson, K. (1964), "Starch-Gel Electrophoresis – Application to the Classification of Pituitary Proteins and Polypeptides", *Metabolism* **13**, 985-1002.

[89] Serwer, P., Allen, J. L., and Hayes, S. J. (1983), "Dependence of Gel Sieving on the Agarose Preparation", *Electrophoresis* **4**, 232-235.

[90] Wieme, R. J. (1965), "Agar Gel Electrophoresis", Elsevier, Amsterdam, The Netherlands.

[91] Arnott, S. Fulmer, A., Scott, W. E., Dea, I. C. M., Moorehouse, R., and Rees, D. A. (1974), "The Agarose Double Helix and Its Function in Agarose Gel Structure", *J. Mol. Biol.* **90**, 269-284.

[92] Cook, R. B., and Witt, H. J. (1981), "Agarose Composition, Aqueous Gel and Method of Making Same", *US Patent* No. 4290911.

[93] Hansson, H. A., and Kagedal, S. L. (1982), "Medium for Isoelectric Focusing", *US Patent* No. 4312739.

[94] Cook, R. B. (1982), "Derivatized Agarose and Method of Making and Using Same", *US Patent* No. 4319975.

[95] Ghosh, S., and Moss, D. B. (1974), "Electroendosmosis Correction for Electrophoretic Mobility Determined in Gels", *Anal. Biochem.* **62**, 365-370.

[96] Serwer, P., and Hayes, S. J. (1982), "Correction of Electrophoretic Mobilities for the Electro-Osmosis of Agarose", *Electrophoresis* **3**, 80-84.

[97] Buzas, Zs., and Chrambach, A. (1982), "Steady-State Stacking in Agarose at Various pH", *Electrophoresis* **3**, 121-129.

[98] Stellwagen, N. C. (1983), "Accurate Molecular Weight Determinations of DNA Restriction Fragments on Agarose Gels", *Biochemistry* **22**, 6180-6185.

[99] Ben-Or, S., and Chrambach, A. (1983), "Heterogeneity of the Glucocorticoid Receptors: Molecular Transformations During Activation, Detected by Electrofocusing", *Arch. Biochem. Biophys.* **221**, 343-353.

[100] Salokangas, A., Eppenberger, U., and Chrambach, A. (1981), "Isolation of Active cAMP Dependent Protein Kinases from Calf Ovaries Gel Electrophoresis *vs* Gel Electrofocusing", *Preparative Biochem.* **12**, 299-320.

[101] Chrambach, A., Reisfeld, R. A., Wyckoff, M., and Zaccari, J. (1967), "A Rapid and Sensitive Method for the Staining of Proteins Fractionated on Polyacrylamide Gels", *Anal. Biochem.* **20**, 150-154.

[102] Diezel, W., Kopperschläger, G., and Hofmann, E. (1972), "An Improved Procedure for Protein Staining in Polyacrylamide Gels with a New Type of Coomassie Brilliant Blue", *Anal. Biochem.* **48**, 617-620.

[103] Axelsen, N. H., Kroll, J., and Weeke, B. (1973), *A Manual of Quantitative Immunoelectrophoresis*, pp. 1-169. Universitetsforlaget, Oslo.

[104] Fenner, C., Traut, R. R., Mason, D. T., and Wikman-Coffelt, J. (1975), "Quantification of Coomassie Blue Stained Proteins in Polyacrylamide Gels Based on Analyses of Eluted Dye", *Anal. Biochem.* **63**, 595-602.

[105] An der Lan, B., Cowell, J. L., Burstyn, D. G., Manclark, C. R., and Chrambach, A. (1985), "Characterization of the Filamentous Hemagglutinin from Bordetella Pertussis by Gel Electrophoresis", *Archives Biochem. Biophys.*, submitted.

[106] Dirksen, M. L., and Chrambach, A. (1972), "Studies on the REDOX State in Polyacrylamide Gels", *Separation Sci.* **7**, 747-772.

[107] Bennick, A. (1968), "On the Spectrophotofluorometric Determinationof Nanogram Amounts of α,ε,-Diaminopimelic Acid, a Bacteria Cell Wall Constituent", *Anal. Biochem.* **26**, 457-458.

[108] Rodbard, D., and Chrambach, A. (1974), "Quantitative Polyacrylamide Gel Electrophoresis (PAGE): Mathematical and Statistical Analysis of PAGE Data", in: *Electrophoresis and Electrofocusing on Polyacrylamide Gel*, pp. 28-62. (Allen, R. C., and Maurer, H. R., eds.). W. de Gruyter, Berlin-New York NY.

[109] Lang, U., Kahn, C. R., and Chrambach, A. (1980), "Characterization of the Insulin Receptor and Insulin Degrading Activity from Human Lymphocytes by Quantitative Polyacrylamide Gel Electrophoresis", *Endocrinology* **106**, 40-49.

[110] Houghten, R., and Chrambach, A. (1979), "Studies on the Reaction Between Human Growth Hormone and Oleic Acid, Using Polyacrylamide Gel Electrophoresis", *Arch. Biochem. Biophys.* **197**, 163-169.

[111] Ben-Or, S., and Chrambach, A. (1984), "The Glucocorticoid Receptors of the Neural Retina of the Chick Embryo: (I) Dependence of Molecular Size and Charge of Its Components on Exposure of the Cell-Free Cytosol to Heat, Salt and Molybdate", in preparation.

[112] Yadley, R. A., Rodbard, D., and Chrambach, A. (1973), "Isohormones of Human Growth Hormone (III): Isolation by Preparative Polyacrylamide Gel Electrophoresis and Characterization", *Endocrinology* **93**, 866-873.

[113] Rodbard, D. (1976), "Estimation of Molecular Weight by Gel Filtration and Gel Electrophoresis", in: *Methods of Protein Separation*, pp. 145-218. (Catsimpoolas, N., ed.), Vol.2, Plenum Press, New York NY.

[114] Sluyterman, L. A. Ae., Elgersma, O., and Wijdenes, J. (1978), "Chromatofocusing: Isoelectric Focusing on Ion Exchange Columns", *J. Chromatogr.* **150**, 17-44.

[115] Buzas, Zs., Hjelmeland, L. M., and Chrambach, A. (1983), "Formation of Natural pH Gradients in Sequential Moving Boundary Systems with Solvent Counterions (II): Experimental Results", *Electrophoresis* **4**, 27-35.

[116] Chrambach, A., and Nguyen, N. Y. (1977), "Electrofocusing in Buffers: Formation of Natural pH Gradients, Flexibility, Gradient Stability, Relation to Isotachophoresis and Preparative Potential", in: *Electrofocusing and Isotachophoresis*, pp. 51-58. (Radola, B. J., and Graesslin, D., eds.). W. de Gruyter, Berlin-New York NY.

[117] Bjellqvist, B., Ek, K., Righetti, P. G., Gianazza, E., Görg, A., Westermeier, R., and Postel, W. (1982), "Isoelectric Focusing in Immobilized pH Gradients: Principle, Methodology and Some Applications", *J. Biochem. Biophys. Methods* **6**, 317-340.

[118] Nguyen, N.Y., and Chrambach, A. (1977), "Natural pH Gradients in Buffer mixtures: Formation in the Abscence of Strongly Acidic and Basic Anolyte and Catholyte, Gradient Steepening by Sucrose and Stabilization by High Buffer Concentrations in the Electrolyte Chambers", *Anal. Biochem.* **79**, 462-469.

[119] An der Lan, B., and Chrambach, A. (1980), "Modification of pH Gradients in Electrofocusing by the pH of the Anolyte", *Electrophoresis* **1**, 23-27.

[120] Ben-Or, S., Sullivan, J. V., and Chrambach, A. (1982), "Voltage Regulation Across Multiple Gel Tubes in Electrofocusing, Using Weakly Acidic and Basic Anolytes and Catholytes", in: *Electrophoresis '82*, pp. 109-120 (Stathakos, D., ed.). W. de Gruyter, Berlin-New York NY.

[121] McCormick, A., Miles, L. E. M., and Chrambach, A. (1976), "Selective Elution of Zones by Ampholytes or Buffers: A Method for Preparative Electrofocusing on Polyacrylamide Gel", *Anal. Biochem.* **75**, 314-324.

[122] Saravis, C. A., and Zamcheck, N. (1979), "Isoelectric Focusing in Agarose", *J. Immunol. Methods* **29**, 91-96.

[123] Radola, B. J. (1976), "Isoelectric Focusing in Granulated Gels", in: *Isoelectric Focusing*, pp. 119-172 (Catsimpoolas, N., ed.). Academic Press, New York NY.

[124] Jackiw, B. A., and Brown, R. K. (1980), "Field Strength Changes During Isoelectric Focusing", *Electrophoresis* **1**, 107-112.

[125] Jackiw, B. A., Chidakel, B. E., Chrambach, A., and Brown, R. K. (1980), "A Device for Measuring the Segmental Voltages Along Electrofocusing Gels", *Electrophoresis* **1**, 102-106.

[126] Nguyen, N. Y., and Chrambach, A. (1976), "Non-isoelectric Focusing in Buffers", *Anal. Biochem.* **74**, 145-153

[127] Bier, M., Palusinski, O. A., Mosher, R. A., and Saville, D. A. (1983), "Electrophoresis: Mathematical Modeling and Computer Simulation", *Science* **219**, 1281-1287.

[128] Svensson, H. (1962), "Isoelectric Fractionation, Analysis and Characterization of Ampholytes in Natural pH Gradients. II: Buffering Capacity and Conductance of Isoionic Ampholytes", *Acta Chem. Scand.* **16**, 456-466.

[129] Righetti, P. G., Gianazza, E., Dossi, G., Celentano, F., Bjellqvist, B, Sahlin, B., and Eklund, C. (1984), "Generation of Highly-Reproducible, Extended pH Intervals in Immobiline Gels", in: *Electrophoresis '83*, (Hirai, H., ed.). W. de Gruyter, Berlin-New York NY, pp. 533-540.

[130] Nguyen, N. Y., and Chrambach, A. (1980), "Electrofocusing on Flat pH Gradients: Systematic pH Gradient Modification Leading to Improved Protein Separation", *Electrophoresis* **1**, 14-22.

[131] Nguyen, N. Y., Salokangas, A., and Chrambach, A. (1977), "Electrofocusing in Natural pH Gradients Formed by Buffers: Gradient Modification and Stability with Time", *Anal. Biochem.* **78**, 287-294.

[132] Prestidge, R. L., and Hearn, M. T. W. (1981), "The Application of Buffer Electrofocusing to Granulated Flat Bed Media", *J. Separation Purif. Methods* **10**, 1-28.

[133] Hearn, M. T. W., Prestidge, R. L., Griffin, J. F. T., and Mhlanga, G. W. (1981), "Purification of Pregnancy-Associated α-Macroglobulin by Buffer Electrofocusing", *Preparative Biochem.* **11**, 191-200.

[134] Vinogradov, S. N., Lowenkron, S., Andonian, M. R., Bagshaw, J., Felgenhauer, K., and Pak, S. J. (1973), "Synthetic Ampholytes for the Isoelectric Focusing of Proteins", *Biochem. Biophys. Res. Commun.* **54**, 501-506.

[135] Righetti, P. G. (1983), "Isoelectric Focusing: Theory, Methodology and Applications", pp. 1-36. Elsevier, Amsterdam, The Netherlands.

[136] Gelsema, W. J., De Ligny, C. L., and Van der Veen, N. G. (1979), "Comparison of the Specific Conductivities, Buffer Capacities and Molecular Weights of Focused Ampholine, Servalyte and Pharmalyte Carrier Ampholytes Used in Isoelectric Focusing", *J. Chromatogr.* **173**, 33-41.

[137] Cuono, C., Chapo, G., Chrambach, A., and Hjelmeland, L. M. (1983), "Detection of Binding Artifacts by Use of a Polymeric Dye in Gel Electrofocusing in Simple Buffers and Commercial Synthetic Carrier Ampholyte Mixtures", *Electrophoresis* **4**, 404-407.

[138] Righetti, P. G., and Gianazza, E. (1978), "Isoelectric Focusing of Heparin Evidence for Complexing with Carrier Ampholytes", *Biochim. Biophys. Acta* **532**, 137-146.

[139] Gianazza, E., Astorri, C., and Righetti, P. G. (1979), "Ampholine-Ampholine Interaction as a Cause of pH Gradient Drift in Isoelectric Focusing", *J. Chromatogr.* **171**, 161-169.

[140] Vesterberg, O. (1969), "Synthesis and Isoelectric Fractionation of Carrier Ampholytes", *Acta Chem. Scand.* **23**, 2653-2666.

[141] Arosio, P., Gianazza, E., and Righetti, P. G. (1978), "Coexistence of Steady State and Transient State in Isoelectric Focusing", *J. Chromatography* **166**, 55-64.

[142] Baumann, G., and Chrambach, A. (1975), "Mechanisms of pH Gradient Instability in Electrofocusing", in: *Progress in Isoelectric Focusing and Isotachophoresis*, pp. 13-23 (Righetti, P. G., ed.). Elsevier, Amsterdam, The Netherlands.

[143] Fawcett, J. S. (1975), "The pH Drift During Isoelectric Focusing: A Comparison between Different Techniques", in: *Progress in Isoelectric Focusing and Isotachophoresis*, pp. 25-37 (Righetti, P. G. ed.). Elsevier, Amsterdam, The Netherlands.

[144] McCormick, A. G., Wachslicht, H., and Chrambach, A. (1978), "Separation of Hemoglobins A and S by Gel Electrofocusing, Using Selective Zone Elution by Gel Transposition between Suitable Anolytes and Catholytes", *Anal. Biochem.* **85**, 209-218.

[145] Righetti, P. G., and Macelloni, C. (1982), "New Polyacrylamide Matrices for Drift-Free Isoelectric Focusing", *J. Biochem. Biophys. Methods* **6**, 1-16.

[146] Swanson, M. J., and Sanders, B. E. (1975), "Isoelectric Focusing in Polyacrylamide Gel Using Buffer Electrode Solutions", *Anal. Biochem.* **67**, 520-524.

[147] Finlayson, R., and Chrambach, A. (1971), "Electrofocusing in Polyacrylamide Gel and Its Preparative Application", *Anal. Biochem.* **40**, 292-311.

[148] Catsimpoolas, N. (1975), "Transient State Isoelectric Focusing: Toward the Steady State and Beyond", in: *Progress in Isoelectric Focusing and Isotachophoresis*, pp. 77-92 (Righetti, P. G., ed.). Elsevier, Sci. Publ. Co., Amsterdam, The Netherlands.

[149] Caspers, M. L., and Chrambach, A. (1977), "Natural pH Gradients Formed By Aminoacids: Ampholyte Distribution, Time Course, Use in Electrofocusing of Protein, Relation to pH Gradients in Isotachophoresis, Separator Effects", *Anal. Biochem.* **81**, 28-39.

[150] Gianazza, E., Righetti, P. G., Bordi, S., and Papeschi, G. (1977), "pH Determinations in Isoelectric Focusing with an Iridium Electrode", in: *Electrofocusing and Isotachophoresis* , pp. 173-180 (Radola, B. J. and Graesslin, D., eds.). W. de Gruyter, Berlin-New York NY.

[151] Reisner, A. H., Nemes, P., and Bucholtz, C. (1975), "The Use of Coomassie Briliant Blue G-250 Perchloric Acid Solution for Staining in Electrophoresis and Isoelectric Focusing on Polyacrylamide Gels", *Anal. Biochem.* **64**, 509-516.

[152] Vesterberg, O., Hansen,L., and Sjosten A. (1977), "Staining of Proteins after Isoelectric Focusing in Gels by New Procedures", *Biochim. Biophys. Acta* **491**, 160-166.

[153] Boddin, M., Hilderson, H. J., Lagrou, A., and Dierick, W. (1975), "Preparative Isoelectric Focusing: Elution of the Gradient Without Interrupting the Electrical Field", *Anal. Biochem.* **64**, 293-296.

[154] Ben-David, M., and Chrambach A. (1977), "Preparation of Bio-and Immunoactive Human Prolactin in Milligram Amounts from Amniotic Fluid in 60% Yield", *Endocrinology* **101**, 250-261.

[155] Nguyen, N. Y., Grindeland, R. E., and Chrambach, A. (1980), "Isolation of Human Growth Hormone Isohormones D and E in Milligram Amounts (II), Using Electrofocusing on Polyacrylamide Gel", *Preparative Biochem.* **11**, 173-189.

[156] Nwokoro, N., Chen, H.-C., and Chrambach, A. (1981), "Physical, Biological and Immunological Characterization of Highly Purified Urinary Human Chorionic Gonadotropin Components Separated by Gel Electrofocusing", *Endocrinology* **108**, 291-300.

[157] Baumann, G., and Chrambach, A. (1975), "Quantitative Removal of Carrier Ampholytes from Protein Fractions Derived from Electrofocusing", *Anal. Biochem.* **69**, 649-651.

[158] Horuk, R., Beckner, S., Lin, M., Wright, D. E., and Chrambach, A. (1984), "Purification of the Photoaffinity-Labeled Glucagon Receptor by Gel Electrophoretic Methods", *Preparative Biochem.* **14**, 99-121.

[159] Nguyen, N. Y., and Chrambach, A. (1979), "Amino Acid Spacing in Isotachophoresis on Polyacrylamide Gel: A Critical Evaluation", *Anal. Biochem.* **94**, 202-210.

[160] Nguyen, N. Y., Baumann G., Arbegast D. E., Grindeland, R. E., and Chrambach, A. (1980), "Isolation of Human Growth Hormone Isohormones D and E in Milligram Amounts (I), Using Isotachophoresis on Polyacrylamide Gel", *Preparative Biochem.* **11**, 139-157.

[161] Wachslicht, H., and Chrambach, A. (1978), "A Simple Device for Protein Concentration by Steady-State Stacking", *Anal. Biochem.* **84**, 533-538.

[162] Thormann, W. (1983), "Description and Detection of Moving Sample Zones in Zone Electrophoresis: Zone Spreading due to the Sample as a Necessary Discontinuous Element." *Electrophoresis* **4**, 383-389.

[163] Zoon, K. C., Smith, M. E., Bridgen, P. J., Zur Neden, D., and Anfinsen, C. B. (1979), "Purification and Partial Characterization of Human Lymphoblastoid Interferon", *Proc. Natl. Acad. Sci. USA* **76**, 5601-5605.

[164] Bosshard, H. F., and Datyner, A. (1977), "The Use of a New Reactive Dye for Quantitation of Prestained Proteins on Polyacrylamide Gels", *Anal. Biochem.* **82**, 327-333.

[165] Mardian, J. K. W., and Isenberg, I. (1978), "Preparative Gel Electrophoresis: Detection, Excision and Elution of Protein Bands from Unstained Gels", *Anal. Biochem.* **91**, 1-12.

[166] Nguyen, N. Y., DiFonzo, J., and Chrambach, A. (1980), "Protein Recovery from Gel Slices by Steady-State Stacking. An Apparatus for the Simultaneous Extraction and Concentration of Ten Samples", *Anal. Biochem.* **106**, 78-91.

[167] Jovin, T., Chrambach, A., and Naughton, M. A. (1964), "An Apparatus for Preparative Temperature Regulated Polyacrylamide Gel Electrophoresis", *Anal. Biochem.* **9**, 351-369.

[168] Nees, S. (1974), in: *Electrophoresis and Electrofocusing in Polyacrylamide Gel*, pp. 189-198 (Allen, R. C., and Maurer, H. R., eds.). W. de Gruyter, Berlin-New York NY.

[169] Chen, B., Chrambach, A., and Rodbard, D. (1979), "Continuous Optical Scanning in Polyacrylamide Gel Electrophoresis: Estimation of the Apparent Diffusion Coefficient of β-lactoglobulin B", *Anal. Biochem.* **97**, 120-130.

[170] Deutsch, D. G. (1976), "Effect of Prolonged 100 °C Treatment in Sodium Dodecyl Sulfate upon Peptide Bond Cleavage", *Anal. Biochem.* **71**, 300-303.

[171] Steinhardt, J., and Fugitt, C. H. (1942), "Catalyzed Hydrolysis of Amide and Peptide Bonds in Proteins", *J. of Res. Natl. Bureau Standards* **29**, 315-327.

[172] Nguyen, N. Y., and Chrambach, A. (1980), "Protein Composition of Plasmin Preparation 'Homolysin'", *Preparative Biochem.* **11**, 159-172.

[173] O'Farrell, P. H. (1975), "High Resolution Two-Dimensional Electrophoresis of Proteins", *J. Biol. Chem.* **250**, 4007-4021.

[174] Chrambach, A., Allen, R. C., Klose, J., Merril, C. R., Righetti, P. G., and Kronberg, H. (1983), "Roundtable Discussion on Accuracy and Precision in 2-Dimensional Macromolecular Mapping", in: *Electrophoresis '82*, pp. 797-825 (Stathakos, D., ed.). W. de Gruyter, Berlin-New York NY.

[175] Poehling, H. M., and Neuhoff, V. (1980), "One and Two-Dimentional Electrophoresis in Micro-Slab Gels", *Electrophoresis* **1**, 90-102.

[176] Nguyen, N. Y., Rodbard, D., Svendsen, P. J., and Chrambach, A. (1977), "Cascade Stacking and Cascade Electrofocusing. Their Interconversion and Fundamental Unity", *Anal. Biochem.* **77**, 39-55.

[177] Switzer, R. C., Merril, C. R., and Shifrin, S. (1979), "A Highly Sensitive Silver Stain for Detecting Proteins and Peptides in Polyacrylamide Gels", *Anal. Biochem.* **98**, 231-237.

[178] Vincent, R. K., Hartman, J., Barrett, A. S., and Sammons, D. W. (1981), "Multispectral Digital Image Analysis of Color Two-Dimensional Electrophoretograms", in: *Electrophoresis '81*, pp. 371-381 (Allen, R. C., and Arnaud, P., eds.). W. de Gruyter, Berlin-New York, NY.

[179] Kronberg, H., Zimmer, H. G., and Neuhoff, V. (1980), "Photometric Evaluation of Slab Gels. I. Data Acquisition an Image Analysis", *Electrophoresis* **1**, 27-31.

[180] Lemkin, P., and Lipkin, L.E. (1981), "Gellab: Multiple 2D Electrophoretic Gel Analysis", in: *Electrophoresis '81*, pp. 401-411 (Allen, R. C., and Arnaud, P., eds.). W. de Gruyter, Berlin-New York NY.

[181] Klose, J., and Schneider, W. (1980), "Gegenwärtiger Stand und Trends in der Entwicklung der Zweidimensionalen Elektrophorese", in: *Elektrophorese Forum '80*, pp. 91-94 (Radola B. J., ed.). Technische Universität München, FRG.

[182] Taylor, J., Anderson, N. L., and Anderson, N. G. (1981), "A Computerized System for Matching and Stretching Two-Dimensional Gel Patterns Represented by Parameter Lists", in: *Electrophoresis '81*, pp. 383-400 (Allen, R. C., and Arnaud, P., eds.). W. de Gruyter, Berlin-New York NY.

[183] Merril, C. R., Goldman, D., and Ebert, M. (1981), "Quantitative Two-Dimensional Electrophoresis as a Screen for Genetic Disease Markers", in: *Electrophoresis '81*, p. 343-354 (Allen, R. C., and Arnaud, P., eds.). W. de Gruyter, Berlin-New York NY.

[184] Chrambach, A., Cantz, M., and Kapadia, G. (1972), "Isotachophoresis in Polyacrylamide Gel", *Separation Sci.* **7**, 785-816.

[185] Bui, C., Galea, V., and Chrambach, A. (1977), "An Apparatus for the Simultaneous Performance of Polyacrylamide Gel Electrophoresis at Multiple pH", *Anal. Biochem.* **81**, 108-117.

[186] Skyler, J. S., Baumann, G., and Chrambach, A. (1977), "A Catalogue of Isohormones of Human Growth Hormone Based on Quantitative Polyacrylamide Gel Electrophoresis", *Acta Endocr. Suppl.* **211**, 1-40.

[187] Tipton, H., Rumen N.M., and Chrambach, A., (1975), "A Slicer for Polyacrylamide Gels of 3-, 6-, and 18 mm Diameter", *Anal. Biochem.* **69**, 323-326.

[188] Horuk, R., and Wright, D. E. (1983), "Partial Purification and Charactrization of the Glucagon Receptor", *FEBS Lett.* **155**, 213-217.

[189] Duke J. A., Bier, M., and Nord, F. F. (1952), "On the Mechanism of Enzyme Action III: The Amphoteric Properties of Trypsin", *Arch. Biochem. Biophys.* **40**, 424-430.

[190] Cannan , R. K., Kibrick, A., and Palmer, A. H. (1941), "The Amphoteric Properties of Egg Albumin", *Ann. N.Y. Acad. Sci.* **41**, 243-266.

[191] Nguyen, N. Y., and Chrambach, A. (1977), "Stabilization of pH Gradients in Buffer Electrofocusing on Polyacrylamide Gel", *Anal. Biochem.* **82**, 54-62.

[192] Nguyen, N. Y., and Chrambach, A. (1977), "Stabilization of pH Gradients Formed by Ampholine", *Anal. Biochem.* **82** , 226-235.

[193] Righetti, P. G., and Chrambach, A. (1978), "Hindered Migration of Aminoacids Toward Their Isoelectric Positions in Gel Electrofocusing, and Facilitation of the Migration at High Ionic Strength", *Anal. Biochem.* **90**, 633-643.

[194] Frey, M. D., Atta, M. B., and Radola, B. J. (1984), "Preparation of Rehydratable Polyacrylamide Gels and Their Application in Ultrathin-Layer Isoelectric Focusing", in: *Electrophoresis '84*, pp. 122-125 (Neuhoff, V., ed.) Verlag Chemie, Weinheim-Deerfield Beach, Florida-Basel.

[195] Chrambach, A. (1984), "Recent Developments in Buffer Electrofocusing", in: *Electrophoresis '84*, pp. 3-28 (Neuhoff, V., ed.) Verlag Chemie, Weinheim-Deerfield Beach, Florida-Basel.

[196] Bell, P. H., McClintock, D. K., and Snedeker, E. H. (1975), "Slicer and Scanner for Polyacrylamide Gel Blocks", *Anal. Biochem.* **65** , 586-590.

[197] Auzan, C., Menard, J., Corvol, P., and Chrambach, A. (1985) "Gel Electrophoretic Separation of Angiotensins I, II and III. Separation and Purification Methods", in press.

Appendixes

Appendix 1

Selected Moving Boundary Electrophoresis Buffer Systems

The following is a survey on MBE buffer systems providing stacking of proteins at various pHs, 0 °C. The selection of systems is made on the basis of low values for the trailing ion mobility.

Buffer system number 1

Catholyte: 2.06 g GABA + 10.0 mL 1 N KOH/L $pH_{25°C}$ = 10.52
Anolyte: 2.10 g Ammediol + 10.0 mL 1 N HCl/L; $pH_{25°C}$ = 8.83

Trailing phase $pH_{0°C}$	Trailing ion mobility	Leading phase					Trailing phase		
		Leading ion Hydrochloric acid (HCl) [M]	Common ion Ammediol [M]	Leading ion 1.0 N HCl [ml/100 mL] (4×)	Common ion Ammediol [g/100 mL] (4×)	$pH_{25°C}$	Trailing ion γ-Amino-butyric acid (GABA) [M]	Common ion Ammediol [M]	$pH_{25°C}$
10.50	0.064	0.1094	0.1289	43.74	5.42	8.20	0.0776	0.0971	9.73
10.77	0.119	0.0652	0.1913	26.08	8.05	9.22	0.0463	0.1724	10.00
10.99	0.174	0.0446	0.2951	17.84	12.41	9.66	0.0317	0.2821	10.23
11.27	0.256	0.0302	0.5328	12.08	22.41	10.12	0.0214	0.5240	10.51

Buffer system number 2

Catholyte: As system 1
Anolyte: 2.42 g Tris + 10.0 mL 1 N HCl/L; $pH_{25°C}$ = 8.07

Trailing phase $pH_{0°C}$	Trailing ion mobility	Leading phase					Trailing phase		
		Leading ion HCl [M]	Common ion Tris [M]	Leading ion 1 N HCl [mL/100 mL] (4×)	Common ion Tris [g/100 mL] (4×)	$pH_{25°C}$	Trailing ion GABA [M]	Common ion Tris [M]	$pH_{25°C}$
10.32	0.048	0.1658	0.3690	66.32	17.87	8.29	0.1189	0.3221	9.52
10.59	0.085	0.0904	0.5987	36.16	29.00	8.93	0.0649	0.5732	9.80
10.74	0.112	0.0683	0.8222	27.32	39.83	9.21	0.0490	0.8029	9.95

Buffer system number 3

Catholyte	1.50 g Glycine + 10.0 mL 1 N KOH/L; pH$_{25°C}$ = 9.74
Anolyte	2.98 g TEA + 10.0 mL 1 N HCl/L; pH$_{25°C}$ = 7.80

Trailing phase pH$_{0°C}$	Trailing ion mobility	Leading phase — Leading ion HCl (M)	Common ion Triethanolamine (TEA) (M)	Leading ion 1.0 N HCl (mL/100 mL)	4 × Common ion TEA (g/100 mL)	pH$_{25°C}$	Trailing phase — Trailing ion Glycine (M)	Common ion TEA (M)	pH$_{25°C}$
9.41	0.060	0.1611	0.1627	64.44	9.71	5.93	0.1237	0.1253	8.77
9.64	0.097	0.0998	0.2299	39.92	13.72	8.03	0.0766	0.2067	9.01
9.94	0.171	0.0567	0.4144	22.68	24.73	8.69	0.0435	0.4012	9.32
10.16	0.245	0.0396	0.6631	15.84	39.57	9.08	0.0304	0.6539	9.54

Buffer system number 4

Catholyte	As system 3
Anolyte	As system 3

Trailing phase pH$_{0°C}$	Trailing ion mobility	Leading phase — Leading ion Acetic acid (M)	Common ion TEA (M)	Leading ion Acetic Acid (g/100 mL)	4 × Common ion TEA (g/100 mL)	pH$_{25°C}$	Trailing phase — Trailing ion Glycine (M)	Common ion TEA (M)	pH$_{25°C}$
9.30	0.048	0.1619	0.1049	3.89	6.26	4.87	0.1575	0.1005	8.66
9.55	0.081	0.0950	0.1746	2.28	10.42	7.84	0.0924	0.1720	8.92
9.88	0.155	0.0498	0.3597	1.20	21.47	8.68	0.0485	0.3584	9.26
10.21	0.266	0.0290	0.7481	0.70	44.64	9.26	0.0282	0.7473	9.60

Buffer system number 5

Catholyte	2.50 g Taurine + 10.0 mL 1 N KOH/L, pH$_{25°C}$ = 9.02
Anolyte	2.31 g NEM + 10.0 mL 1 N HCl/L·pH$_{25°C}$ = 7.69

Trailing phase pH$_{0°C}$	Trailing ion mobility	Leading phase – Leading ion: Lactic acid (M)	Leading phase – Common ion: N-Ethyl-morpholine (M)	Leading phase 4× – Leading ion: Lactic acid (g/100 mL)	Leading phase 4× – Common ion: NEM (g/100 mL)	pH$_{25°C}$	Trailing phase – Trailing ion: Taurine (M)	Trailing phase – Common ion: NEM (M)	pH$_{25°C}$
8.63	0.048	0.1266	0.0532	4.56	2.45	3.57	0.1218	0.0484	8.11
8.98	0.098	0.0626	0.0979	2.26	4.51	7.44	0.0602	0.0955	8.48
9.26	0.159	0.0385	0.1695	1.39	7.81	8.30	0.0370	0.1680	8.76
9.54	0.250	0.0244	0.3238	0.88	14.92	8.84	0.0235	0.3229	9.08

Buffer system number 6

Catholyte	As system 5
Anolyte	As system 5

Trailing phase pH$_{0°C}$	Trailing ion mobility	Leading phase – Leading ion: Acetic acid (M)	Leading phase – Common ion: NEM (M)	Leading phase 4× – Leading ion: Acetic acid (g/100 mL)	Leading phase 4× – Common ion: NEM (g/100 mL)	pH$_{25°C}$	Trailing phase – Trailing ion: Taurine (M)	Trailing phase – Common ion: NEM (M)	pH$_{25°C}$
8.65	0.050	0.1276	0.0619	3.06	2.85	4.60	0.1129	0.0472	8.12
8.98	0.098	0.0619	0.0940	1.49	4.33	7.50	0.0548	0.0869	8.48
9.25	0.159	0.0381	0.1573	0.92	7.25	8.26	0.0337	0.1529	8.76
9.54	0.250	0.0242	0.2966	0.58	13.66	8.80	0.0214	0.2938	9.08

Buffer system number 7

| | | Catholyte | | 1.24 g Boric acid + 10 mL 1 N KOH/L; pH$_{25°C}$ = 9.20 |
| | | Anolyte | | 1.36 g Imidazole + 10 mL 1 N HCl/L; pH$_{25°C}$ = 6.99 |

Trailing phase pH$_{0°C}$	Trailing ion mobility	Leading phase Leading ion Lactic acid (M)	Leading phase Common ion Imidazole (M)	4× Leading ion Lactic acid (g/100 mL)	4× Common ion Imidazole (g/100 mL)	pH$_{25°C}$	Trailing phase Trailing ion Boric acid (M)	Trailing phase Common ion NEM (M)	pH$_{25°C}$
8.40	0.051	0.1284	0.1009	4.63	2.75	4.25	0.1249	0.0974	8.04
8.74	0.101	0.0644	0.2026	2.32	5.52	7.42	0.0626	0.2008	8.37
9.01	0.164	0.0396	0.3630	1.43	9.89	7.98	0.0385	0.3619	8.63
9.30	0.258	0.0251	0.7067	0.90	19.24	8.49	0.0244	0.7060	8.91

Buffer system number 8

| | | Catholyte | | As system 7 |
| | | Anolyte | | 2.3 Lg NEM + 10 mL 1 N HCl/L; pH$_{25}$ °C = 7.69 |

Trailing phase pH$_{0°C}$	Trailing ion mobility	Leading phase Leading ion Acetic acid (M)	Leading phase Common ion NEM (M)	4× Leading ion Acetic acid (g/100 mL)	4× Common ion NEM (g/100 mL)	pH$_{25°C}$	Trailing phase Trailing ion Boric acid (M)	Trailing phase Common ion Imidazole (M)	pH$_{25°C}$
8.32	0.042	0.1395	0.0387	3.35	1.78	4.22	0.1255	0.0247	8.02
8.55	0.069	0.0850	0.0449	2.04	2.07	4.68	0.0765	0.0364	8.25
8.88	0.132	0.0445	0.0731	1.07	3.37	7.58	0.0400	0.0686	8.58
9.30	0.258	0.0228	0.1681	0.55	7.74	8.55	0.0205	0.1658	8.99

Buffer system number 9

Trailing phase pH$_{0°C}$	Trailing ion mobility	Leading phase					Trailing phase			Catholyte	Anolyte
		Leading ion	Common ion	Leading ion (4×)	Common ion (4×)		Trailing ion	Common ion		3.58 g Tricine + 10.0 mL 1 N KOH/L; pH$_{25°C}$ = 8.0	As system 8
		Lactic acid	NEM	Lactic acid	NEM	pH$_{25°C}$	Tricine	NEM	pH$_{25°C}$		
		M		g/100 mL			M				
7.88	0.051	0.0947	0.0492	3.41	2.27	3.75	0.0625	0.0170	7.39		
8.25	0.099	0.0488	0.0432	1.76	1.99	4.60	0.0322	0.0266	7.80		
8.53	0.147	0.0329	0.0527	1.19	2.43	7.54	0.0217	0.0415	8.10		
9.23	0.259	0.0187	0.1744	0.67	8.03	8.66	0.0123	0.1680	8.86		

Buffer system number 10

Trailing phase pH$_{0°C}$	Trailing ion mobility	Leading phase					Trailing phase			Catholyte	Anolyte
		Leading ion	Common ion	Leading ion (4×)	Common ion (4×)		Trailing ion	Common ion		As system 9	As system 2
		Acetic acid	Tris	Acetic acid	Tris	pH$_{25°C}$	Tricine	Tris	pH$_{25°C}$		
		M		g/100 mL			M				
7.88	0.051	0.0944	0.0427	2.27	2.07	4.55	0.0629	0.0112	7.30		
8.25	0.099	0.0487	0.0289	1.17	1.40	4.81	0.0324	0.0126	7.65		
8.53	0.148	0.0328	0.0260	0.79	1.26	5.24	0.0218	0.0150	7.90		
9.23	0.259	0.0187	0.0410	0.45	1.99	8.20	0.0125	0.0348	8.51		

Buffer system number 11

Catholyte	4.58 g TES + 10.0 mL 1 N KOH/L; $pH_{25°C}$ = 7.44
Anolyte	2.62 g HEM + 10.0 mL 1 N HCl/L; $pH_{25°C}$ = 6.89

Trailing phase $pH_{0°C}$	Trailing ion mobility	Leading phase						Trailing phase		
		Leading ion Lactic acid (M)	Common ion Hydroxyethyl morpholine (HEM) (M)		Leading ion Lactic acid (g/100 mL)	4× Common ion HEM (g/100 mL)	$pH_{25°C}$	Trailing ion TES (M)	Common ion HEM (M)	$pH_{25°C}$
7.05	0.045	0.1163	0.0435		4.19	2.28	3.49	0.0900	0.0172	6.60
7.38	0.086	0.0611	0.0393		2.20	2.06	3.97	0.0473	0.0255	6.96
7.71	0.148	0.0356	0.0508		1.28	2.67	6.60	0.0275	0.0427	7.31
8.15	0.250	0.0210	0.1060		0.76	5.56	7.55	0.0162	0.1012	7.79

Buffer system number 12

Catholyte	As system 11
Anolyte	4.18 g Bistris + 10.0 mL 1 N HCl/L; $pH_{25°C}$ = 6.50

Trailing phase $pH_{0°C}$	Trailing ion mobility	Leading phase						Trailing phase		
		Leading ion Cacodylic acid (M)	Common ion Bistris (M)		Leading ion Cacodylic acid (g/100 mL)	4× Common ion Bistris (g/100 mL)	$pH_{25°C}$	Trailing ion TES (M)	Common ion Bistris (M)	$pH_{25°C}$
7.05	0.045	0.1016	0.0358		5.61	3.00	5.78	0.0905	0.0247	6.59
7.39	0.086	0.0530	0.0477		2.93	3.99	6.29	0.0472	0.0419	6.94
7.61	0.127	0.0360	0.0677		1.99	5.67	6.71	0.0321	0.0638	7.17
8.15	0.250	0.0163	0.1972		1.01	16.67	7.61	0.0183	0.1992	7.74

Buffer system number 13

Catholyte	3.64 g ACES + 10.0 mL 1 N KOH/L; $pH_{25°C}$ = 6.80
Anolyte	1.58 g Pyridine + 10.0 mL 1 N HCL/L; $pH_{25°C}$ = 5.21

Trailing phase $pH_{0°C}$	Trailing ion mobility	Leading ion HCl (M)	Common ion Pyridine (M)	4× Leading ion 1 N HCl (mL/100 mL)	4× Common ion Pyridine (g/100 mL)	$pH_{25°C}$	Trailing ion ACES (M)	Common ion Pyridine (M)	$pH_{25°C}$
6.42	0.051	0.1709	0.1726	68.36	5.46	3.36	0.0923	0.0940	6.01
6.76	0.098	0.0888	0.2329	35.56	7.37	5.53	0.0480	0.1920	6.36
6.99	0.145	0.0600	0.3460	24.00	10.95	5.98	0.0324	0.3184	6.60
7.27	0.215	0.0404	0.6139	16.16	19.42	6.44	0.0218	0.5953	6.89

Buffer system number 14

Catholyte	3.68 g HFAH + 10.0 mL 1 N KOH/L; $pH_{25°C}$ = 6.80
Anolyte	As system 13

Trailing phase $pH_{0°C}$	Trailing ion mobility	Leading ion Lactic acid (M)	Common ion Pyridine (M)	4× Leading ion Lactic acid (g/100 mL)	4× Common ion Pyridine (g/100 mL)	$pH_{25°C}$	Trailing ion Hexafluoroacetone hydrate (HFAH) (M)	Common ion Pyridine (M)	$pH_{25°C}$
5.81	0.050	0.1132	0.0525	4.08	1.66	3.63	0.0909	0.0302	5.73
6.14	0.097	0.0589	0.0654	2.12	2.07	4.67	0.0473	0.0538	6.04
6.47	0.166	0.0344	0.1094	1.24	3.46	5.63	0.0276	0.1026	6.34
6.82	0.258	0.0221	0.2241	0.80	7.09	6.23	0.0177	0.2197	6.66

Buffer system number 15

		Catholyte	1.20 g Acetic acid + 10.0 mL 1 N KOH/L; $pH_{25°C}$ = 4.72
		Anolyte	1.50 g Glycine + 10.0 mL 1 N HCl/L; $pH_{25°C}$ = 2.39

Trailing phase $pH_{0°C}$	Trailing ion mobility	Leading phase					Trailing phase		
		Leading ion KOH	Common ion Acetic acid	Leading ion 1 N KOH	4× Common ion Acetic acid	$pH_{25°C}$	Trailing ion Glycine	Common ion Acetic acid	$pH_{25°C}$
		M		mL/100 mL	g/100 mL		M		
3.50	0.046	0.1867	0.2558	74.68	6.14	5.01	0.1147	0.1838	3.45
3.17	0.090	0.0960	0.4186	38.40	10.05	4.08	0.0590	0.3816	3.12
2.91	0.143	0.0604	0.7049	24.16	16.93	3.60	0.0371	0.6816	2.86
2.71	0.196	0.0440	1.0937	17.60	26.27	3.26	0.0270	1.0767	2.66

Buffer system number 16

		Anolyte	As system 15
		Catholyte	As system 15

Trailing phase $pH_{0°C}$	Trailing ion mobility	Leading phase					Trailing phase		
		Leading ion Pyridine	Common ion Acetic acid	Leading ion Pyridine	4× Common ion Acetic acid	$pH_{25°C}$	Trailing ion Glycine	Common ion Acetic acid	$pH_{25°C}$
		M		g/100 mL	g/100 mL		M		
3.50	0.046	0.1489	0.2180	4.71	5.24	4.69	0.1147	0.1838	3.45
3.20	0.085	0.0811	0.3752	2.57	9.01	4.03	0.0625	0.3566	3.15
2.93	0.138	0.0499	0.6606	1.58	15.87	3.54	0.0384	0.6491	2.88
2.73	0.191	0.0360	1.0416	1.14	25.02	3.20	0.0277	1.0333	2.68

Buffer system number 17

		Anolyte				As system 15			
		Catholyte				1.78 g β-Alanine + 10.0 mL 1 N HCl/L; pH$_{25°C}$ = 3.59			
		Leading phase (4 ×)					Trailing phase		
Trailing phase pH$_{0°C}$	Trailing ion mobility	Leading ion Pyridine (M)	Common ion Acetic acid (M)	Leading ion Pyridine (g/100 mL)	Common ion Acetic acid (g/100 mL)	pH$_{25°C}$	Trailing ion β-Alanine (M)	Common ion Acetic acid (M)	pH$_{25°C}$
4.75	0.048	0.1570	0.0488	4.97	1.17	5.69	0.1288	0.0206	4.68
4.41	0.094	0.0762	0.0451	2.41	1.08	5.31	0.0314	0.0314	4.35
4.14	0.153	0.0469	0.0578	1.48	1.39	4.83	0.0385	0.0494	4.09
3.94	0.212	0.0399	0.0792	1.07	1.90	4.52	0.0278	0.0731	3.89

Buffer system number 18

		Anolyte				2.76 g Cacodylic acid + 10.0 mL 1 N KOH/L; pH$_{25°C}$ = 6.21			
		Catholyte				2.06 g GABA + 10.0 mL 1 N HCl/L; pH$_{25°C}$ = 4.07			
		Leading phase (4 ×)					Trailing phase		
Trailing phase pH$_{0°C}$	Trailing ion mobility	Leading ion Imidazole (M)	Common ion Cacodylic acid (M)	Leading ion Imidazole (g/100 mL)	Common ion Cacodylic acid (g/100 mL)	pH$_{25°C}$	Trailing ion γ-Amino-butyric acid (GABA) (M)	Common ion Cacodylic acid (M)	pH$_{25°C}$
5.22	0.050	0.1634	0.1365	4.45	7.53	6.76	0.1382	0.1113	5.19
4.85	0.102	0.0774	0.2567	2.11	14.17	5.72	0.0655	0.2448	4.83
4.71	0.134	0.0589	0.3466	1.60	19.13	5.42	0.0498	0.3375	4.68
4.38	0.230	0.0344	0.7075	0.94	39.05	4.84	0.0291	0.7022	4.36

Buffer system number 19

Anolyte	3.64 g ACES + 10.0 mL 1 N KOH/L; $pH_{25°C} = 6.80$
Catholyte	1.58 g Pyridine + 10.0 mL 1 N HCl/L; $pH_{25°C} = 3.59$

Trailing phase $pH_{0°C}$	Trailing ion mobility	Leading phase					Trailing phase		
				4 ×					
		Leading ion	Common ion	Leading ion	Common ion	$pH_{25°C}$	Trailing ion	Common ion	$pH_{25°C}$
		Imidazole	ACES	Imidazole	ACES		Pyridine	ACES	
		M		g/100 mL			M		
6.70	0.052	0.1736	0.0611	4.73	4.45	7.45	0.1666	0.0541	6.29
6.41	0.095	0.0947	0.0993	2.58	7.24	6.87	0.0909	0.0955	5.99
6.22	0.138	0.0651	0.1444	1.77	10.52	6.47	0.0625	0.1418	5.80
5.85	0.267	0.0336	0.3222	0.91	23.48	5.77	0.0323	0.3209	5.41

Table 1. Other Discontinuous Buffer Systems Reported in the Literature

Polarity	Resolving gel Trailing phase pH[a]	Stacking gel		Ref.No.	Notes[b]
		Trailing phase pH[a]	Trailing ion mobility[a]		
−		10.45	0.064	[97]	P
		9.27	0.045		
		8.10	0.067		
		7.81	0.045		
		7.10	0.049		
		6.42	0.052		
+		5.15	0.056		
		4.71	0.052		
		3.39	0.058		
−	5.85	5.15	0.044	[2]	P
	7.18	6.32	0.041		
	7.95	7.08	0.041		
	8.53	7.82	0.045		
	10.25	9.63	0.096		
	10.88	10.22	0.040		
+	2.82	3.55	0.046		
	4.00	4.88	0.035		
	5.95	6.41	0.094		
	6.33	7.38	0.042		
	8.48	9.00	0.060		
	9.00	10.00	0.027		
−	5.57	5.00	0.031	[67]	
−		10.24	0.041	[52]	P
		9.33	0.005		
		10.13	0.033		
		10.42	0.060		
		10.61	0.088		
−		6.12	0.096	[42]	P
		6.43	0.052		
		6.77	0.025		
		7.10	0.049		
		7.42	0.093		
		7.81	0.045		
		8.07	0.068		
		9.63	0.096		
		10.13	0.033		
		10.45	0.064		
+		4.88	0.035		
		4.62	0.062		
		4.21	0.137		
		3.84	0.024		
		3.67	0.036		
		3.16	0.092		
		3.10	0.114		

Table 1. Other Discontinuous Buffer Systems Reported in the Literature

Polarity	Resolving gel Trailing phase pH[a]	Stacking gel		Ref.No.	Notes[b]
		Trailing phase pH[a]	Trailing ion mobility[a]		
−	7.00	6.12	0.096	[17]	
	7.00	6.43	0.052		
	7.50	7.10	0.049		
	7.50	7.42	0.093		
	8.50	7.81	0.045		
	8.50	8.10	0.056		
	10.25	9.63	0.096		
	11.00	10.45	0.064		
−	7.50	7.38	0.086	[186]	P
	7.50	6.77	0.025		
+	4.14	4.62	0.062		
−	7.79	7.39	0.086		
+	3.50	4.62	0.062	[74]	
	3.67	3.67	0.301		
−	5.48	4.23	0.186		
	7.99	7.11	0.051		
−	7.50	7.09	0.049	[184]	
−		6.12	0.096	[185]	P
		6.43	0.052		
		7.10	0.049		
		7.42	0.093		
		6.77	0.025		
		7.81	0.045		
		7.97	0.099		
		8.10	0.056		
		9.63	0.096		
		10.45	0.064		
− (SDS)	9.58[c]	9.39[c]	0.224[c]	[78]	P

[a] all pH values and mobilities refer to 0 °C
[b] P partial output
[c] 25 °C

Appendix 2

The Extensive Moving Boundary Electrophoresis Buffer Systems Output

A. Selection of a buffer system from the catalogue

It is assumed that the desired resolving pH and the desired polarity of migration and temperature are known. The latter 2 parameters determine the quadrant of System Numbers of interest in the catalogue according to Table 1, App. 2. The system numbers (column 2, Fig. 2) in each quadrant are given in increasing order of pH and in 0.5 pH intervals. The 0.5 pH unit section approximating most closely the operative pH of interest is selected. This operative pH is given as pH(9) on each page (column 9, Fig. 2). It is also given as pH (9) in the heading section of each 0.5 pH unit block of systems (the heading section is not shown in Fig. 2). The most important criterion in selecting the optimal buffer system among those with appropriate pH(9) is the trailing ion mobility [RM (1,4)] in the stacking phase (column 11 of Fig. 2). A value of about 0.050 is desirable for proteins in general. If several buffer systems with low values of RM(1,4) are available, selection is then continued on the basis of the desirability of buffer constituents 1, 2, 3 and 6 for each system (defined by Table 2). Compatibility with the protein, solubility, commercial value and availability usually governs that selection. Another criterion of system's selection is the stacking pH associated with any one pH(9). It is given as pH(4) in column 7 of Fig. 2. Finally, the range of variability of each pH(4) and RM(1,4) given in the line below each system number may be of importance in the selection of the optimal system.

B. Location of the selected buffer system in the microfiche output

The microfiche output provides the properties of 4269 buffer systems in order of their system number. Three pages are devoted to each buffer system. Page 1 reiterates the input parameters by which the system was generated (not shown here). Page 2 tabulates the physico-chemical properties of the system (top section, exemplified by Fig. 3) and gives an explicit recipe for preparation of upper, stacking gel and resolving gel buffers (bottom section, exemplified by Fig. 4). Page 3 provides the composition of the lower buffer (Phase EPSILON) in the top section (not shown here), and tabulates the possible subsystems in order to RM(1,4) or RM(1,9) (Fig. 5). The microfiche output of 3 pages for the selected buffer system number is located in numerical order, using a microfiche reader.

C. Inspection of the physical properties of the buffer system

Page 2 (top) lists the physical properties of the system (Fig. 3). It restates and defines Constituents 1, 2, 3 and 6 (Fig. 3, top), the trailing ion and leading ion mobilities in the stacking phase [RM(1,ZETA) and RM(2,BETA)] and the trailing ion mobility in the resolving phase [RM(1, PI)]. It lists pH and ionic strength of the operative stacking phase (ZETA) and resolving phase (PI). If the values of RM (1,ZETA), RM(2,BETA) and RM(1, PI) do not satisfy one's requirements, a subsystem with the desired values is selected as described under E. If the ionic strength for Phase ZETA is undesirable, the concentration of the Constituents C2 and C6 of Phase BETA are multiplied by the factor desired ION.STR.(ZE-TA)/listed ION.STR.(ZETA) to obtain the desired stacking gel buffer, as described in D.

D. Preparation of buffers

Page 2 (bottom) lists the buffer recipes for all but the lower buffer (Phase EPSI-LON). The latter is given on p.3 (top) as 0.0625 M Constituent 6, and 0.05 M HCl or KOH, respectively. Note that pH and specific conductance (μmhos/cm) are given for 25 °C to alleviate measurement. Gel buffers are given in 4-fold final concentration to allow for preparation of polyacrylamide gels by the mixing ratio 2:1:1 (Fig. 25).

If one wants to alter the ionic strength of Phases ZETA or PI, the specified amounts of constituents of the corresponding setting phases 4× BETA or 4× GAMMA are multiplied by the factor of the desired ionic strength over the ionic strength of the phase in the table of properties (Fig. 3 and Section C).

E. Preparation of buffers for a subsystem

Page 3 of the output (center section) lists the subsystems in order of RM (1,ZETA) or RM(1, PI) (Fig. 5). The desired value of RM(1,ZETA) or RM(1, PI) is selected from column 1. Columns 8 and 9 in the same line then give the final molar concentrations of Constituents 2 and 6 for a desired value of RM(1,ZETA). The columns 11 and 12 give the final molar concentrations of Constituents 3 and 6 for a desired value of RM(1, PI).

An easy way of preparing the buffers consists of multiplying the amounts (Fig. 4) by the ratio of the final concentration in the subsystem (Fig. 5) over the final concentration in the numbered system (Fig. 3, C2–C6 or C3–C6).

F. Computation of the properties of a subsystem

The physical properties of a subsystem are computed by the program of T. M. Jovin using the input instructions in Appendix 3 and the input format shown in Fig. 6. The final molar concentrations for Phases ZETA (here used as upper buffer), BETA and GAMMA for the desired subsystem (see E) are transcribed from the table shown in Fig. 5 to line 64 of the input. Column 60 specifies (left to right, Fig. 6) the date, operator initials and system number (up to 5 digits), shown in Fig. 6. Line 63 defines Constituents 1, 2, 3, 4, 5 and 6. Columns 1 and 2 in line 61 define polarity and temperature. Columns 2 and 3 in line 65 relate to elution buffer (in elution PAGE) and are put at pH(PI) – 1 and +1, respectively. All other input parameters may remain constant as shown in Fig. 6. The listed order of final molar concentrations in line 64 then is, from left to right, that of Constituents (1, 6), (2, 6) and (3, 6). For our purpose of computing the ionic strength of a stacking gel, it is of course not necessary to list any resolving gel concentrations. Line 64 would then give the final molar concentrations of Constituents (1, 6), (2, 6) and (2, 6).

The reader may not want by himself to locate stacking systems at reasonable pH intervals, low RM(1,ZETA) and constant ionic strength of 0.01 M by himself, but rather rely on those systems previously reported in the literature (Table 1 of App. 1).

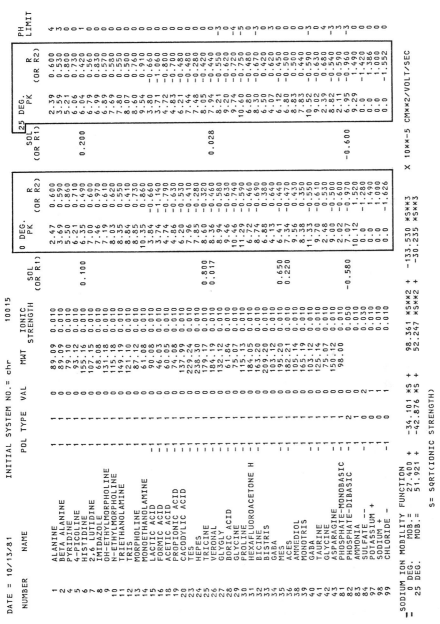

Figure 1. Buffer constituents used to generate the "Extensive Buffer Systems Output". Each constituent is characterized by values, both at 0 and 25 °C, of ionic mobility relative to Na^+ (R) and pK.

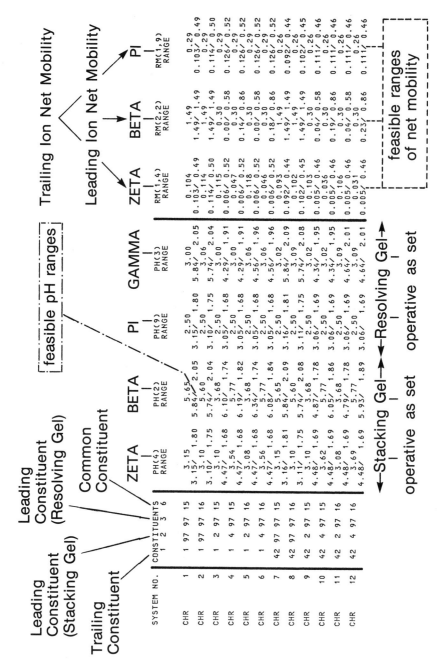

Figure 2. Representative format of the MBE buffer systems catalog (NTIS No. PB-196 090).

SYSTEM NUMBER

```
DATE = 03/09/84     COMPUTER SYSTEM NUMBER = chr      10013
POLARITY = + (MIGRATION TOWARD CATHODE)   TEMPERATURE =   0  DEG. C.
```

```
CONSTITUENT 1 = NO.  42 , GLYCINE
CONSTITUENT 2 = NO.  34 , GABA
CONSTITUENT 3 = NO.  34 , GABA
CONSTITUENT 6 = NO.  15 , LACTIC ACID
```

PHASES

	ALPHA(1)	ZETA(4)	BETA(2)	PI(9)	LAMBDA(8)	GAMMA(3)
C1	0.1600	0.1599		0.0625		
C2			0.1768		0.0691	
C3						0.0691
C6	0.0235	0.0235	0.0404	0.0536	0.0602	0.0602
THETA	0.147	0.147	0.229	0.858	0.871	0.871
PHI(1)	0.060	0.060		0.160		
PHI(2)			0.202		0.542	
PHI(3)						0.542
PHI(6)	0.407	0.407	0.885	0.187	0.622	0.622
RM(1)	0.032	0.032		0.085		
RM(2)			0.129		0.347	
RM(3)						0.347
RM(6)	-0.269	-0.268	-0.584	-0.123	-0.411	-0.411
PH	3.68	3.68	4.73	3.20	4.06	4.06
ION.STR.	0.0096	0.0096	0.0358	0.0100	0.0375	0.0375
SIGMA	1.098	1.100	4.485	1.148	4.699	4.699
KAPPA	273.	274.	1057.	285.	1105.	1105.
NU	0.029	0.029	0.029	0.074	0.074	0.074
BV	0.034	0.034	0.075	0.038	0.072	0.072

Figure 3. Representative section of page 2 of the "Extensive Buffer Systems Output" for a MBE buffer system providing the physical-chemical properties of that system. Trailing and leading ion net mobilities, pH and ionic strength of each buffer phase are marked as being of particular practical importance.

RECIPES FOR BUFFERS OF PHASES ZETA(4),BETA(2),GAMMA(3),PI(9)

		Upper Buffer	Stacking Gel Buffer	Resolving Gel Buffer	
		1X	4X	4X	4X
CONSTITUENT		PHASE 4	PHASE 2	PHASE 3	PHASE 9
GLYCINE	GM	12.01			1.88
GABA	GM		7.29		
GABA	GM			2.85	
LACTIC ACID	GM	2.12	1.46	2.17	1.93
H2O TO		1 LITER	100 ML	100 ML	100 ML
AT FINAL CONCENTRATION =					
PH(25 DEG.C.)		3.61	4.67	4.01	3.13
KAPPA(25 DEG.C.)		573.	1999.	2080.	594.

Figure 4. Representative section of page 2 of the "Extensive Buffer Systems Output" for a MBE buffer system providing the composition of upper, stacking gel and resolving gel buffers.

Subsystem PI ZETA

I —1—

X —10—

Stacking Gel or Resolving Gel					Stacking Gel					Resolving Gel		
STACKING AND UNSTACKING RANGES												
PHASE ZETA(4) OR PI(9)					PHASE BETA(2) OR LAMBDA(8)					PHASE GAMMA(3)		
RM(1)	PHI(1)	C(1)	C(6)	PH	RM(2)	PHI(2)	C(2)	C(6)	PH	C(3)	C(6)	PH
0.005	0.010	0.1599	0.0020	4.48	0.06	0.101	0.1768	0.0188	5.08	1.1052	0.1177	5.08
0.032	0.060	0.1599	0.0236	3.67	0.13	0.203	0.1768	0.0405	4.73	0.1842	0.0422	4.73
0.058	0.110	0.1599	0.0674	3.39	0.24	0.367	0.1768	0.0843	4.37	0.1005	0.0479	4.37
0.085	0.160	0.1599	0.1373	3.20	0.35	0.542	0.1768	0.1541	4.06	0.0691	0.0602	4.06
0.111	0.210	0.1599	0.2381	3.06	0.44	0.684	0.1768	0.2550	3.80	0.0526	0.0759	3.80
0.138	0.260	0.1599	0.3763	2.93	0.50	0.782	0.1768	0.3932	3.57	0.0425	0.0945	3.57
0.164	0.310	0.1599	0.5599	2.83	0.54	0.847	0.1768	0.5768	3.39	0.0357	0.1163	3.39
0.191	0.360	0.1599	0.7996	2.73	0.57	0.891	0.1768	0.8164	3.22	0.0307	0.1418	3.22
0.217	0.410	0.1599	1.1096	2.64	0.59	0.920	0.1768	1.1264	3.07	0.0270	0.1717	3.07
0.244	0.460	0.1599	1.5094	2.55	0.60	0.941	0.1768	1.5262	2.93	0.0240	0.2074	2.93
0.270	0.510	0.1599	2.0266	2.46	0.61	0.956	0.1768	2.0434	2.79	0.0217	0.2505	2.79
0.297	0.560	0.1599	2.7011	2.38	0.62	0.967	0.1768	2.7180	2.67	0.0197	0.3034	2.67
0.323	0.610	0.1599	3.5935	2.29	0.62	0.975	0.1768	3.6104	2.54	0.0181	0.3700	2.54
0.350	0.660	0.1599	4.8000	2.19	0.63	0.981	0.1768	4.8168	2.41	0.0167	0.4562	2.41

Trailing Ion Net Mobility

Leading Ion Net Mobility

Trailing Phase pH

Final Molar Concentrations

Figure 5. Representative section of page 3 of the "Extensive Buffer Systems Output" for a MBE buffer system, providing the subsystems of indentical constituent composition but varying in constituent concentrations. The table provides trailing and leading ion net mobilities and concentrations of stacking gel and resolving gel buffers.

```
 1.    //ACC999J  JOB  (JFV1,675,A),ANDREAS.CHRAMBACH
 2.    //SKR111S2 EXEC FORVCALL,REGION.GO=300K,NAME='JFV1MLD.JOVINLIB',
 3.    //  DISK=FILE22,PROGRAM=MLDJOVIN
 4.    //GO.FT01F001 DD *
 5.       45    7
 6.      1 ALANINE                    +1 1 4    89.09 0.01          2.47  0.600      2.39  0
 7.      2 BETA ALANINE               +1 1 3    89.09 0.01          3.69  0.590      3.59  0
 8.      4 PYRIDINE                   +1 1 0    79.10 0.01          5.50  0.86       5.21  0
 9.      5 4-PICOLINE                 +1 1 0    93.12 0.01          6.21  0.71       6.06  0
10.      6 HISTIDINE                  +1 1+1   155.16 0.01   0.1    6.35  0.49  0.2  6.04  0
11.      7 2,6 LUTIDINE               +1 1 0   107.15 0.01          7.00  0.60       6.79  0
12.      8 IMIDAZOLE                  +1 1 0    68.08 0.01          7.46  0.97       6.99  0
13.      9 OH-ETHYLMORPHOLINE         +1 1 0   131.18 0.01          7.19  0.61       6.89  0
14.     10 N-ETHYLMORPHOLINE          +1 1 0   115.18 0.01          8.03  0.62       7.69  0
15.     11 TRIETHANOLAMINE            +1 1 0   149.19 0.01          8.35  0.55       7.80  0
16.     12 TRIS                       +1 1 0   121.10 0.01          8.84  0.41       8.07  0
17.     13 MORPHOLINE                 +1 1 0    87.12 0.01          8.85  0.73       8.60  0
18.     14 MONOETHANOLAMINE           +1 1 0    61.08 0.01         10.35  0.86       9.54  0
19.     15 LACTIC ACID                -1 1 0    90.08 0.01          3.84 -0.66       3.80 -0
20.     16 FORMIC ACID                -1 1 0    46.03 0.01          3.74 -1.14       3.71 -1
21.     18 ACETIC ACID                -1 1 0    60.05 0.01          4.74 -0.79       4.72 -0
22.     19 PROPIONIC ACID             -1 1 0    74.08 0.01          4.86 -0.63       4.83 -0
23.     20 CACODYLIC ACID             -1 1 0   137.99 0.01          6.20 -0.53       6.21 -0
24.     23 TES                        -1 1 0   229.24 0.01          7.96 -0.41       7.44 -0
25.     24 HEPES                      -1 1 0   238.3  0.01          7.85 -0.28       7.48 -0
26.     25 TRICINE                    -1 1 0   179.17 0.01 0.8      8.60 -0.32       8.05 -0
27.     26 VERONAL                    -1 1 0   184.19 0.01 0.017    8.36 -0.46 0.028 7.94 -0
28.     27 GLYGLY                     -1 1-3   132.12 0.01          8.94 -0.58       8.21 -0
29.     28 BORIC ACID                 -1 1 0    61.84 0.01          9.46 -0.63       9.20 -0
30.     29 GLYCINE                    -1 1-4    75.07 0.01         10.46 -0.74       9.74 -0
31.     30 PROLINE                    -1 1-5   115.13 0.01         11.29 -0.59      10.60 -0
32.     31 HEXAFLUOROACETONE H        -1 1 0   184.05 0.01          6.72 -0.46       6.80 -0
33.     32 BICINE                     -1 1-3   163.20 0.01          8.74 -0.69       8.30 -0
34.     33 BISTRIS                    +1 1 0   209.20 0.01          6.88  0.38       6.50  0
35.     34 GABA                       +1 1+3   103.12 0.01          4.13  0.64       4.07  0
36.     35 MES                        -1 1 0   195.2  0.01 0.65     6.41 -0.44       6.12 -0
37.     36 ACES                       -1 1     182.21 0.01 0.22     7.34 -0.47       6.80 -0
38.     38 AMMEDIOL                    1 1     105.14 0.01          9.56  0.43       8.83  0
39.     39 MONOTRIS                    1 1     165.19 0.01          8.38  0.35       7.83  0
40.     40 GABA                       -1 1-3   103.12 0.01         11.33 -0.55      10.52 -0
41.     41 TAURINE                    -1 1     125.14 0.01          9.70 -0.61       9.02 -0
42.     42 GLYCINE                     1 1 4    75.07 0.01          2.48  0.53       2.39  0
43.     43 ASPARAGINE                 -1 1-3   150.12 0.01          9.00 -0.50       8.82 -0
44.     81 PHOSPHATE-MONOBASIC        -1 1 3    98.00 0.01          2.02 -0.60       2.11 -0
45.     82 PHOSPHATE-DIBASIC          -1 2-3         0.05 -0.58     7.07 -0.97 -0.60 6.95 -0
46.     83 AMMONIA                    +1 1 0         0.01         10.12  1.52        9.29  1
47.     84 SULFATE --                 -1 0 2         0.03               -1.28            -1.
48.     97 POTASSIUM +                +1 0 1         0.01                1.490            1.
49.     98 SODIUM +                   +1 0 1         0.01                1.000            1.
50.     99 CHLORIDE -                 -1 0 1         0.01               -1.626           -1.
51.     81    1M PHOSPHORIC ACID
52.     82    1M PHOSPHORIC ACID
53.     83    1N NH4OH
54.     84    1M SULFURIC ACID
55.     97    1N KOH
56.     98    1N NAOH
57.     99    1N HCL
58.    27.4      -34.101   98.361    -133.53   51.921   -42.876   52.247   -30.23
59.       97    83    99    18
60.    03/09/84  chr      10013     1    1
61.    +1  0  3        1                                               ←
62.                                                          4.0       ←
63.       42    34    34    97    97    15
64.       .1600    .0235    .1768    .0404    .0691    .0602           ←
65.       3.0      2.2      4.2            0.05      .8
66.    /*
67.    //GO.FT03F001  DD  SYSOUT=A,DCB=(RECFM=UA,BLKSIZE=133)
68.    //GO.SYSUDUMP DD SYSOUT=A
```

Figure 6. Representative input into the Jovin program (NTIS No. PB-196092) for the re-computation of the full 3-page MBE buffer systems output for a subsystem (the composition of which is defined in the format of Fig. 5 of Appendix 2). Appendix 3 provides instructions for alternative input, in particular that for the computation of new MBE buffer systems. Note that each line in the figure corresponds to a "card" in Appendix 3.

Table 1. PB Numbers for the Buffer Systems from the National Technical Information Service. US Department of Commerce, Springfield VA 22161, USA

Positively charged proteins, 0 °C	259309
Positively charged proteins, 25 °C	259310
Negatively charged proteins, 0 °C	259311
Negatively charged proteins, 25 °C	259312
Catalog of 4269 systems	196090

Table 2. Nomenclature of the Extensive Buffer Systems Output

Constituents: 1	Trailing constituent ("trailing ion") in the upper buffer
2	Leading constituent ("leading ion") in the stacking gel buffer
3	Leading constituent in the resolving gel buffer
6	common constituent of a charge opposite to 1, 2, 3.

Phases: ALPHA	Upper buffer (trailing constituent reservoir)
BETA	Stacking gel buffer as prepared
GAMMA	Resolving gel buffer as prepared
EPSILON	Lower buffer (common constituent reservoir)
ZETA	Operative stacking gel buffer
PI	Operative resolving gel buffer

Phase compositions in terms of constituent numbers:

ALPHA	1, 6
BETA	2, 6
GAMMA	3, 6
EPSILON	6-chloride or 6-Na$^+$
ZETA	1, 6
PI	1, 6
LAMBDA	2, 6

RM (1,ZETA)	Trailing ion mobility in the stacking gel, lowest mobility of a constituent of the stack
(2,BETA)	Leading ion mobility in the stacking gel, highest mobility of a constituent of the stack
(1,PI)	Trailing ion mobility in the resolving gel, highest mobility tolerated if the constituent is to be unstacked
	Correspondingly the lowest mobility of a constituent which is to remain stacked in the resolving phase

Moving boundaries:
ZETA/BETA boundary across constituents 1/2 in the stacking phase
PI/LAMBDA boundary across constituents 1/2 in the resolving phase

The relation between function of a phase, composition of a phase in terms of constituents, phase designation and trailing ion mobilities is given by Table 3.

Table 3. Composition of and Moving Boundaries across the Most Important Buffer Phases Defined by the Extensive Buffer Systems Output

Function of Phase	Constituent		Phase Designation		Net Ion Mobilities	
	Leading	Trailing	Prepared	Operative	Leading	Trailing
	Constituent		Phase			
Stacking of protein	2	1	BETA	ZETA	RM(2,BETA)	RM(1,ZETA)
Resolving (Unstacking) of protein	2	1	GAMMA	PI	RM(2,LAMBDA)	RM(1,PI)
Preceding[a] the protein	3	2	GAMMA	LAMBDA	RM(3,GAMMA)	RM(2,LAMBDA)

where RM = constituent mobility relative to Na^+

[a] The composition of the phase preceding the protein is of no importance to separation and is only given to explain the origin of Phase LAMBDA which is of importance in regulating the resolving Phase PI

The reader is advised *not* to attempt to memorize this confusing array of terms, but rather to allow himself to be guided through the practical application of these terms, using Tables 1 and 2 as a reference.

Appendix 3

Input instructions for the program (NTIS No. 196092) for the computation of multiphasic buffer systems (by T. M. Jovin)

The program can be used in two ways: (1) For the recalculation of the properties of buffer systems when the constituents and constituent concentrations for phases ALPHA, BETA and GAMMA are known. (2) For computation of new buffer systems within desired constraints of temperature, pH, maximal RM(1,4), minimal RM(2,2), maximal pH difference between phases and constituents.

Fig. 6 of Appendix 2 illustrates application (1). Line 5 corresponds to card number 1 (section 1,I). The constituent data (section 2) are entered into lines 6 to 58 into the columns specified by card number 2 of section 1,I. Line 60 corresponds to section 1,II. Line 61 is defined by section 1, III, card number 1, line 62 by card number 2, line 63 by card number 3, line 64 by card number 4, line 65 by card number 5.

In application (2), lines 60 to 65 in Fig. 6, Appendix 2, are replaced by the input format given in Fig. 1, Appendix 3. Line 19 is defined by section 1,II, card number 1. Section 1,III, card number 1 defines the column positions of line 20, card number 2 those of line 21, card number 3 those of line 22, card number 5 those of line 23.

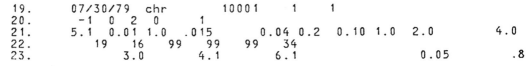

```
19.      07/30/79  chr          10001      1       1
20.       -1   0   2   0          1
21.      5.1   0.01  1.0    .015         0.04  0.2   0.10  1.0    2.0           4.0
22.           19     16     99      99      99     34
23.                 3.0          4.1          6.1                       0.05          .8
```

Figure 1. Input for the computation of new buffer systems by the program of T. M. Jovin.

A. Specification of Input Cards

All entries must be right justified in the field unless otherwise indicated. Numbers in floating-point format can be anywhere in the field. Unused fields must be left blank. Variable names used in the program are indicated in capital letters when necessary. All concentrations are molar.

For numeric data, an asterisk preceding the word "Columns" indicates integer format. The absence of an asterisk indicates floating-point (F) format.

I. Constituent Data

CARD NUMBER 1

*Columns 3–5   NUM (total number of potential constituents for which data supplied. Less than or equal to 100).

*Columns 9–10   NSTAND (number of constituents with identifying numbers greater than 80).

CARD NUMBER 2...NUM + 1

                 A card with the following data is required for each constituent.

*Columns 1–3   Identifying number (less than or equal to 100).

 Columns 5–23   Name (left justified).

*Columns 25–26   Polarity

                 − 1 for anion or base

                 + 1 for cation or acid

*Columns 27–28   IET (electrolyte type)

                 0 for ion

                 1 for monovalent weak electrolyte

                 2 for divalent weak electrolyte

 Columns 29–30   for IET 0,

                 absolute value of ionic valences

                for IET1,2,

                 blank for non-ampholytes

                 for ampholytes (optional);

                 − N if desire to restrict use of constituent to systems with pH(9) no greater than N/pH units *below* its pK

                 + N if desire to restrict use of constituent to systems with pH(9) no greater than N/pH units *above* its pK

Columns 31–38   Molecular weight (for constituents with numbers less than or equal to 80).

Columns 39–43   Ionic strength corresponding to pK and mobility data

Columns 44–49 for IET 0,
           blank
        for IET 1,
           solubility of uncharged acid or base at 0 °C (optional). If not specified no restriction on the basis of solubility is applied.
        for IET 2,
           relative mobility of monovalent ion at 0 °C (signed)

Columns 50–55 for IET 0,
           blank
        for IET 1,2,
           pK at 0 °C

Columns 56–61 for IET 0,1,
           relative ion mobility at 0 °C (signed)
        for IET 2,
           relative mobility of divalent ion at 0 °C (signed)

Columns 62–67 same as for columns 44–49 except for 25 °C
Columns 68–73 same as for columns 50–55 except for 25 °C
Columns 74–79 same as for columns 56–61 except for 25 °C

CARD NUMBER NUM + 2 ... NUM + NSTAND + 2

        A card with the following additional data is required for each constituent with a number greater than 80.

*Columns 1–3   Identifying number
Columns 7–25   Name of constituent in form desired for specification of recipes in terms of 1 M or 1 N solutions (left justified).

CARD NUMBER NUM + NSTAND + 3

        This card specifies the coefficients for the equation giving the sodium ion mobility as a function of $x =$ (ionic strength) 1/2 mobility $= c1 \cdot x + c2 \cdot x^2 + c3 \cdot x^3 + c4 \cdot x^4$ 10^{-5} cm^2/V/s

Columns 1–10   c1 at 0 °C
Columns 11–20 c2 at 0 °C
Columns 21–30 c3 at 0 °C
Columns 31–40 c4 at 0 °C

Columns 41–50 c1 at 25 °C
Columns 51–60 c2 at 25 °C
Columns 61–70 c3 at 25 °C
Columns 71–80 c4 at 25 °C

CARD NUMBER NUM + NSTAND + 4

This card specifies possible *ions* as Constituent 6 in the case of Constituent 2 a monovalent weak electrolyte. NCAT 2 and NAN 2 are used only if their pKs are $\geq$ 4 units from pH(9) so that they are completely ionized. Any or all fields may be blank but the card is required.

*Columns 1–5   NCAT 1 (identifying number of cation)
*Columns 6–10   NCAT 2 (identifying number of monoacidic base)
*Columns 11–15 NAN 1 (identifying number of anion)
*Columns 16–20 NAN 2 (identifying number of monobasic acid)

II. Date, Operator Name, Initial System Number, No. Copies

This card is presented once per run.

Columns 1–8   Date in format 06/07/66
Columns 11–20 Name of operator (left justified)
*Columns 21–25 Initial system number for the given date. In output, columns 11–25 appear as the computer system number, i. e., "NAME X" where X is increased by 1 for each successful system generated. A separate term, SYSTEM NUMBER, is left blank for more permanent and systematic numbering.
*Columns 26–30 Number of copies of systems output desired.
*Columns 31–35 Number of copies of systems summary desired.

III. Data Set For Systems Design

The following series of cards constitute one set of conditions under which all possible systems are generated. Additional data sets are presented sequentially in the deck.

CARD NUMBER 1

*Columns 1–3   SPOL (system polarity)
                      −1 migration to anode
                      +1 migration to cathode

*Columns 4–6   IT (temperature in degrees C.: 0 or 25)

*Columns 7–9   IIN
                1 phase concentrations unspecified. Constituents 1,2,6 unspecified or partially or totally specified. Constituent 3 must be specified.
                2 phase concentrations unspecified. Constituents 1,2,6 unspecified or partially or totally specified. Constituent 3 will be made the same as Constituent 2.
                3 concentrations in Phases 1,2,3 and Constituents 1,2,3,6 specified.

*Columns 10–12 for IIN 1,2,
                IOPT (stacking optimization parameter)
                0 stacking optimized by minimizing absolute value of Constituent 1 mobility in Phase ZETA(4)
                1 stacking optimized by maximizing absolute value of Constituent 2 mobility in Phase BETA(2). Only used for Constituent 2 with IET = 1.
                2 if pH(4) unspecified (i.e., PHZET blank), Constituent 1 mobility in Phase ZETA(4) set = RIZETL pH(4) if pH(4) specified (i.e., PHZET 0), pH(4) set = PHZET
              for IIN 3,
                blank

*Columns 13–15 for INC2 blank (Constituent 2 unspecified), constituents are examined as potential choices for Constituent 2 according to the following table.

ICT2 (Constituent 2 selection parameter)

	ion	weak electrolyte	
		monovalent	divalent
1	+	+	+
2	+		
3		+	
4			+
5	+	+	
6	+		+
7		+	+

for INC2/0 (i.e., Constituent 2 specified), blank

*Columns 16–18 IDE (elution and lower buffer calculation parameter)

0 elution and lower buffers not calculated

1 elution and lower buffers to be calculated

CARD NUMBER 2   for IIN = 3, columns 1–50 are left blank

Columns 1–5   pH(9) [pH during fractionation in Phase PI(9)]

Columns 6–10   R1PIS (lower limit for absolute value of allowed range for RM(1,9), the relative Constituent 1 mobility in Phase PI(9)

Columns 11–15   R1PIL [upper limit for absolute value of allowed range for RM(1,9)]

Columns 16–20   XIS(9) [ionic strength desired for Phase PI(9)]

Columns 21–25   PHZET (pH desired for Phase ZETA(4)-optional)

Columns 26–30   C(1,4) [molar concentration of Constituent 1 in Phase ZETA(4)]

Columns 31–35   R1ZETL [upper limit allowed for absolute value of RM(1,4), relative Constituent 1 mobility in Phase ZETA(4)]

Columns 36–40   R2BETS [lower limit allowed for absolute value of RM(2,2), relative Constituent 2 mobility in Phase BETA(2)]

Columns 41–45 T1 [maximal value allowed for SPOL.
(pH(2)–pH(4))] – if left blank, a value of 4
is automatically used

Columns 46–50 CMAX (maximal allowed molar concentration in system)

Columns 51–55 T2 [maximal value allowed for SPOL.
(pH(7)–pH(9))] – if left blank a value of 4 is
automatically used

Columns 56–60 RFMAX [upper limit allowed for ratio
RM(7,7)/RM(1,9)]

CARD
NUMBER 3

constituent specification

for IIN 1, specification Constituents 1,2,6
optional; Constituent 3 is required

IIN 2, specification Constituents 1,2,6
optional; Constituent 3 is left
blank

IIN 3, specification Constituents 1,2,6,3 required

for IDE 0, Constituents 4 and 5 left blank

IDE 1,Constituents 4 and 5 required

specification of Constituent 7 is always optional

*Columns 1–5   INC1 (identifying number of designated
Constituent 1)

*Columns 6–10   INC2 (identifying number of designated
Constituent 2)

*Columns 11–15 INC3 (identifying number of designated
Constituent 3)

*Columns 16–20 INC4 (identifying number of designated
Constituent 4)

*Columns 21–25 INC5 (identifying number of designated
Constituent 5)

*Columns 26–30 INC6 (identifying number of designated
Constituent 6)

*Columns 31–35 INC7 (identifying number of designated
Constituent 7)

CARD
NUMBER 4

	for IIN 1,2, this card is omitted
	IIN 3, this card is required
Columns 1–10	C(1,1) [concentration of Constituent 1 in Phase ALPHA(1)]
Columns 11–20	C(6,1) [concentration of Constituent 6 in Phase ALPHA(1)]
Columns 21–30	C(2,2) [concentration of Constituent 2 in Phase BETA(2)]
Columns 31–40	C(6,2) [concentration of Constituent 6 in Phase BETA(2)]
Columns 41–50	C(3,3) [concentration of Constituent 3 in Phase GAMMA(3)]
Columns 51–60	C(6,3) [concentration of Constituent 6 in Phase GAMMA(3)]

CARD
NUMBER 5

	for IDE 0, this card is omitted
	IDE 1, this card is required
Columns 1–10	Ratio of ionic strength in elution buffer to that in Phase PI(9)
Columns 11–20	Lower limit for desired range of elution buffer pH
Columns 21–30	Upper limit for desired range of elution buffer pH
Columns 31–40	pH(10) (pH of elution buffer. If specified, columns 11–30 are left blank. Leave blank if range or values is desired).
Columns 41–50	Ionic strength desired for lower buffer
Columns 51–60	Fractional dissociation of Constituent 6 desired in lower buffer

B. Constituent Data Used for Systems Design

Date = 06/26/70   Initial System No. = Jov-Chr   1

Number	Name	Polarity	Type	Val	Molecular mass	Ionic strength	SOL (or R1)	pk at 0°C	R (or R2)	SOL (or R1)	pk at 25°C	R (or R2)	pH Limit
1	Alanine	1	1	0	89.09	0.010		2.47	0.600		2.39	0.600	4
2	Beta Alanine	1	1	0	89.09	0.010		3.69	0.590		3.59	0.530	3
4	Pyridine	1	1	0	79.10	0.010		5.50	0.860		5.21	0.800	0
5	4-Picoline	1	1	0	93.12	0.010		6.21	0.710		6.06	0.730	0
6	Histidine	1	1	0	155.16	0.010	0.100	6.35	0.490	0.200	6.04	0.420	1
7	2,6-Lutidine	1	1	0	107.15	0.010		7.00	0.600		6.79	0.560	0
8	Imidazole	1	1	0	68.08	0.010		7.46	0.970		6.99	0.830	0
9	OH-Ethylmorpholine	1	1	0	131.18	0.010		7.19	0.610		6.89	0.570	0
10	N-Ethylmorpholine	1	1	0	115.18	0.010		8.03	0.620		7.69	0.580	0
11	Triethanolamine	1	1	0	149.19	0.010		8.35	0.550		7.80	0.550	0
12	Tris	1	1	0	121.10	0.010		8.84	0.410		8.07	0.500	0
13	Morpholine	1	1	0	87.12	0.010		8.85	0.730		8.60	0.760	0
14	Monoethanolamine	1	1	0	61.08	0.010		10.35	0.860		9.54	0.910	0
15	Lactic acid	−1	1	0	90.08	0.010		3.84	−0.660		3.80	−0.660	0
16	Formic acid	−1	1	0	46.03	0.010		3.74	−1.140		3.71	−1.060	0
18	Acetic acid	−1	1	0	60.05	0.010		4.74	−0.790		4.72	−0.800	0
19	Propionic acid	−1	1	0	74.08	0.010		4.86	−0.630		4.83	−0.700	0
20	Cacodylic acid	−1	1	0	137.99	0.010		6.20	−0.530		6.21	−0.480	0
23	TES	−1	1	0	229.24	0.010		7.96	−0.410		7.44	−0.480	0
24	HEPES	−1	1	0	238.30	0.010	0.800	7.85	−0.280		7.48	−0.280	0
25	Tricine	−1	1	0	179.17	0.010	0.017	8.60	−0.320		8.05	−0.420	0
26	Veronal	−1	1	0	184.19	0.010		8.36	−0.460	0.028	7.94	−0.440	0
27	Glygly	−1	1	0	132.12	0.010		8.94	−0.580		8.21	−0.550	−3
28	Boric acid	−1	1	0	61.84	0.010		9.46	−0.630		9.20	−0.620	0
29	Glycine	−1	1	0	75.07	0.010		10.46	−0.740		9.74	−0.720	−4
30	Proline	−1	1	0	115.13	0.010		11.29	−0.590		10.60	−0.750	−5
31	Hexafluoroacetone H	−1	1	0	184.05	0.010		6.72	−0.460		6.80	−0.480	0
32	Bicine	−1	1	0	163.20	0.010		8.74	−0.690		8.30	−0.670	−3
33	Bistris	1	1	0	209.20	0.010		6.88	0.380		6.50	0.420	3
34	GABA	1	1	0	103.12	0.010		4.13	0.640		4.07	0.620	3
35	MES	−1	1	0	195.20	0.010	0.650	6.41	−0.440		6.12	−0.450	0
36	ACES	−1	1	0	182.21	0.010	0.220	7.34	−0.470		6.80	−0.500	0
38	Ammediol	1	1	0	105.14	0.010		9.56	0.430		8.83	0.500	0
39	Monotris	1	1	0	165.19	0.010		8.38	0.350		7.83	0.440	0
40	GABA	−1	1	0	103.12	0.010		11.33	−0.550		10.52	−0.590	−3

No.	Name				MW								
41	Taurine	−1	1	0	125.14	0.010		9.70	−0.610		9.02	−0.630	0
42	Glycine	−1	1	0	75.07	0.010		2.48	0.530		2.39	0.680	4
43	Asparagine	−1	1	0	150.12	0.010		9.00	−0.500		8.82	−0.540	−3
81	Phosphate-monobasic	−1	2	0	98.00	0.010		2.02	−0.600		2.11	−0.590	3
82	Phosphate-dibasic	−1	1	0		0.050	−0.580	7.07	−0.970	−0.600	6.95	−0.960	−3
83	Ammonia	1	1	0		0.010		10.12	1.520		9.29	1.490	0
84	Sulfate⁻⁻	−1	0	2		0.030		0.0	−1.280		0.0	−1.420	0
97	Potassium⁺	1	0	1		0.010		0.0	1.490		0.0	1.386	0
98	Sodium⁺	1	0	0		0.010		0.0	1.000		0.0	1.000	0
99	Chloride⁻	−1	0	1		0.010		0.0	−1.626		0.0	−1.552	0

$\times 10^{**-5}$ CM**2/Volt/s

Sodium ion mobility function

0 °C   Mob. = $27.400 + -34.101$ *S + 98.361 *S**2 + -133.530 *S**3

25 °C   Mob. = $51.921 + -42.876$ *S + 52.247 *S**2 + -30.235 *S**3

S = $\sqrt{\text{ionic strength}}$

Appendix 4

Instrumentation

Table 1. Commercial Sources of Instrumentation

Biomed Instruments Inc., 1020 S. Raymond Ave, Suite B, Fullerton CA 92631
Bio-Rad Laboratories, Richmond CA 94804
Bethesda Research Labs. POB 6010, Rockville MD 20850
Brinkmann Instruments Inc., Westbury NY 11590
Buchler Scientific Instruments, 1327 16th Str., Fort Lee NJ 07024
Calbiochem-Behring Corp., POB 12087, San Diego CA 92112
Desaga, POB 101969, D-6900 Heidelberg 1, FRG
E-C Apparatus Corp., 3831 Tyrone Blvd N., St. Petersburg FL 33709
Gelman Sciences, 600 S. Wagner Rd. Ann Arbor, MI 48106
Hoefer Scientific Instruments, Box 77387, San Francisco CA 94107
Ingold Electrodes Inc., I Burtt Road, Andover MA 01810
ISCO, 4700 Superior, Lincoln NE 68504
Isolab Inc., Drawer 4350, Akron OH 44321
Joyce Loebl, Marquisway, Team Valley, Gateshead, Tyne & Wear NE110Q3, UK
LKB Produkter AB, Box 305, S-16126, Bromma, Sweden
Marine Colloids, POB 308, Rockland ME 04841.
Miles Research Products, Elkhart IN 46514
MRA Corp., 1058 Cephas Rd., Clearwater FL 33515
Neslab Instruments, Inc., 871 Islington Str., Portsmouth NH 03801
Ortec Inc., 100 Midland Rd., Cak Ridge TN 37830
Pierce Chemical Co., Box 117, Rockford IL 61105
Pharmacia Fine Chemicals AB, Box 175, S-75104 Uppsala 1, Sweden
Polysciences Inc., Paul Valley Industrial Park, Warrington PA 18976
R & D Scientific Glass Co., 15931 Batson Rd., Spencerville MD 20868
Savant Instruments, 221 Park Ave, Hicksville NY 11801
Serva Fine Biochemicals, PCE A, Garden City Park LI, NY 11040
Shandon Southern Instruments, Inc., 515 Broad Street, Sewickley PA 15143
Sigma Chem. Co., POB 14508, St. Louis MO 63178
A. H. Thomas Co., POB 779, Philadelphia PA 19105

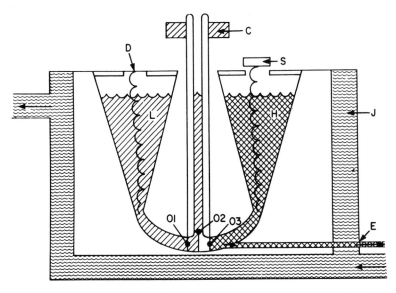

Figure 1. Schematic representation of a polyacrylamide gel gradient maker with temperature control and shielding of the polymerization mixture from inhibition by oxygen through a layer of argon [48]. (C) Control for triple stopcock 01-02-03 (hollow stem); (D) dummy stirring helix; (E) exit port; (H) heavy solution of high %T polymerization mixture; (J) coolant jacket; (L) light solution of low %T polymerization mixture; stopcock openings 01, 02 and 03) allow for selective flow from either L or H into the hollow stem of C, and for flow between L and H, to be able to clear the line between L and H of air bubbles without mixing of solutions L and H.

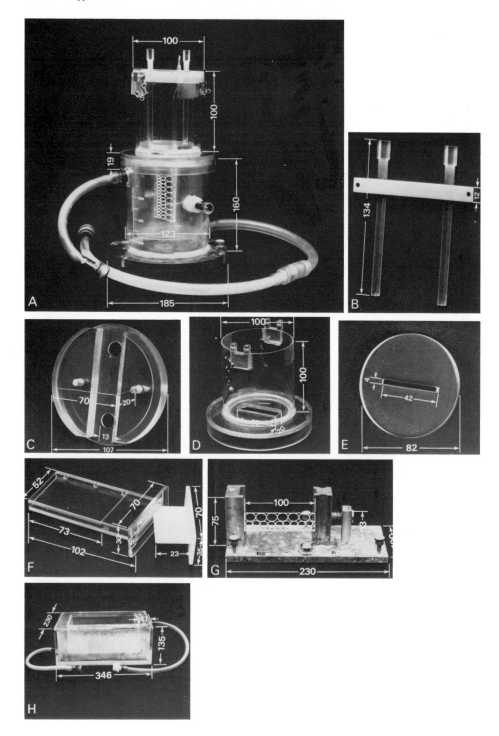

◁ Figure 2. Prototype apparatus for transverse or 2-D longitudinal/transverse pore gradient electrophoresis (Fig. 1 of [17]). (A) Assembly; (B) aluminum bar supporting the Plexiglas rods exerting vertical pressure on the cell; (C) coverplate with upper electrode; (D) upper buffer reservoir with lateral steel pins; (E) Plexiglas pressure plate; (F) cell with slot former; (G) aluminum cell holder; (H) jacketed constant temperature bath for the cell holder.

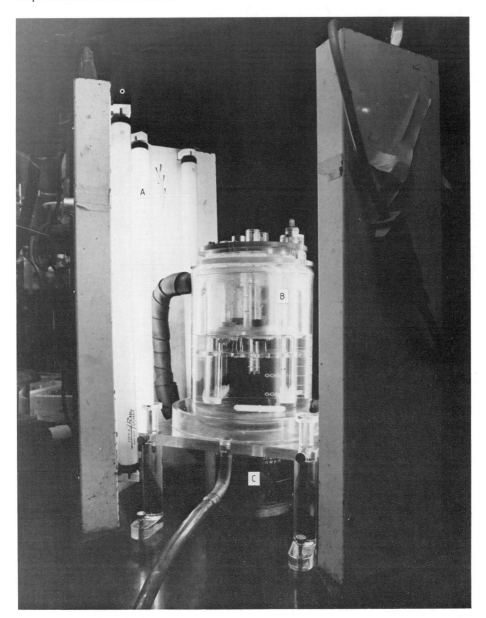

Figure 3. Photopolymerization setup.
(A) Photoilluminators; (B) PAGE apparatus;
(C) Magnetic stirer.

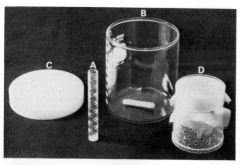

Figure 4. Apparatus for rapid diffusion of SDS from polyacrylamide gels to allow for protein staining in SDS-PAGE within a few hours [55].

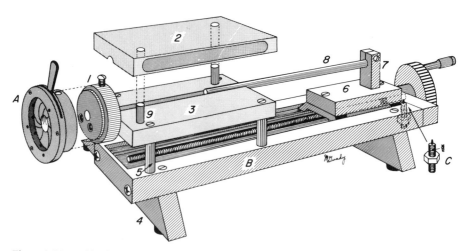

Figure 5. Manual horizontal gel slicer for cylindrical gels of 3, 6 and 18 mm diameter [187], [58]. (A) Cutting iris diaphragm; (B) slide assembly with drivescrew; (C) spring plunger; (1) iris diaphragm holder; (2) piston for 6 mm gels; (3) gel holder bottom plate with groove; (4) support stand; (5) piston for 3 mm gels; (6) mounting plate; (7) ram clamp; (8) ram.

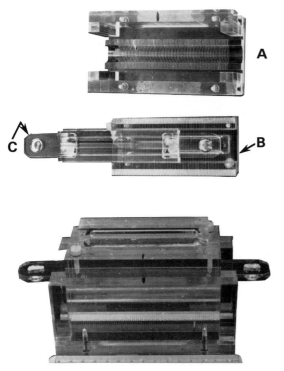

Figure 6. Wire grid transverse gel slicer suitable for 18 mm diameter 5 %T, 15 %C$_{DATD}$ gels. The slicer is a scaled-up version of 10 cm length of the design of [59].

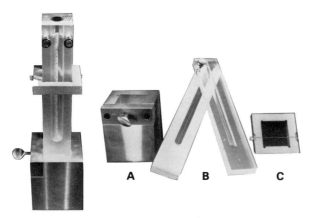

Figure 7. Longitudinal gel slicer for obtaining a 3 mm thick guidestrip of 18 mm diameter preparative gel tubes. The design is a scaled-up version of that reported in [60].

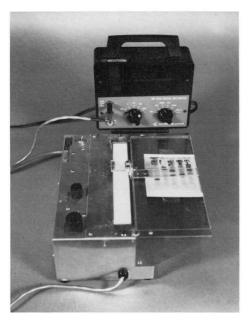

Figure 8. Electronic R_f measuring device [61]. The hairline guide connected to a sliding wire potentiometer is positioned on the origin and moving boundary front, and the digital voltmeter is correspondingly calibrated to 0 and 1.0. Subsequent passage of the hairline along the gel elicits a digital reading of R_f.

Appendix 5

Listing of program IDENT (in BASIC) of D. Rodbard for identity testing between sets of proteins by F-test criteria

```
00010 REM VERSION of 1971; REVISED (F VALUES) 2/7/76
00020 DIMX(70), Y(70), Z(70)
00030 DIMA(70), B(70), C(70), D(70), E(70), F(70), G(70), H(70)
00040 FILE #1, "MUNI"
00050 PRINT "THIS PROGRAM IS EXPERIMENTAL. VALIDITY OF OUT-
      PUT"
00060 PRINT "DEPENDS ON PROPER EXPERIMENTAL DESIGN"
00070 PRINT "TO BE USED IN COLLABORATION WITH D. RODBARD
      M. D."
00080 PRINT "PROGRAM REVISED 2/7/76"
00090 REM A = INTERCEPT B = SLOPE C = XBAR D = YBAR
00100 REM E = N F = SIGMA(W) G = RESVAR H = SUM(W(X-
      XBAR)^2)
00110 PRINT
00120 PRINT "CURVE"; TAB7; "PROTEIN"; TAB15; "D. F."; TAB22;
      "RES.VAR."
00130 FOR I = 1 TO 100
00140 IF END #1, THEN 280
00150 READ #1,X(I)
00160 IF X(I) = 999 THEN 280
00170 READ #1,Y(I), Z(I)
00180 READ #1,A(I), B(I), C(I), D(I)
00190 READ #1, E(I), F(I), G(I), H(I)
00200 PRINT I; TAB10; Z(I); TAB20; E(I)−2; TAB20; G(I)
00210 E=E+E(I)−2
00220 G=G+(E(I)−2)*G(I)
00230 C=C+1/(E(I)−2)
00240 B=B+(E(I)−2)*LOG(G(I))
00250 K=K+1
00260 NEXT I
00270 PRINT
00280 PRINT "G =" G; "E = " E; "MEAN RES. VAR. = G/E ="G/E
00290 C=1+1/3/(K−1)*(C−1/E)
00300 B = E*LOG(G/E)−B
00310 PRINT "B =" B; "C =" C; "CHI SQUARE = B/C =" B/C
00320 PRINT "COMPARE WITH CHI-SQUARE DISTRIBUTION WITH";
      K−1; "D. F."
00330 GOSUB980
00340 PRINT "ENTER YES IF YOU WANT TO USE COMBINED RES.
      VAR. AND D. F."
```

```
00350  PRINT "ENTER NO IF YOU WANT TO ANALYZE EACH CURVE
       SEPARATELY"
00360  INPUT A$
00370  IF A$ = "NO" THEN 690
00380  FOR I = 1 TO K
00390  G(I) = G/E
00400  E(I) = E + 2
00410  NEXT I
00420  PRINT "DO YOU WANT JOINT C.L. ELLIPSES, YES OR NO?"
00430  INPUT B$
00440  IF B$ = "NO" THEN 690
00450  FOR I = 1 TO K
00460  PRINT "CURVE"; I; "SYSTEM" X(I); "C = "Y(I); "PROTEIN";Z(I)
00470  PRINT
00480  V2 = E(I) − 2
00490  GOSUB 1080
00500  D1 = B(I) + SQR(2*G(I)*F/H(I))
00510  D2 = B(I) − SQR(2*G(I)*F/H(I))
00520  PRINT "SLOPE", "YO-LOWER", "YO-UPPER"
00530  FOR D = D2 TO D1 STEP (D1−D2)/4
00540  Z = 2*G(I)*F − (D − B(I))^2*H(I)
00550  Z = Z/F(I)
00560  Z = SQR(Z)
00570  C1 = D(I) −Z
00580  C2 = D(I) +Z
00590  E1 = C1 − D*C(I)
00600  E2 = C2 − D*C(I)
00610  F1 = EXP(E1*LOG(10))
00620  F2 = EXP(E2*LOG(10))
00630  PRINT D,F1,F2
00640  NEXT D
00650  PRINT
00660  PRINT"----------------------"
00670  PRINT
00680  NEXT I
00690  FOR J = 1 TO 100
00700  IF END #1, THEN 970
00710  READ #1,P,Q
00720  PRINT "CONTRAST OF CURVES"; P; "AND"; Q
00730  REM: METHOD OF JOHN GART (PERSONAL COMMUNICA-
       TION)
00740  A1 = A(P)−A(Q)
00750  B1 = B(P)−B(Q)
00760  C1 = 1/F(P) + C(P)^2/H(P) + 1/F(Q) + C(Q)^2/H(Q)
00770  C2 = 1/H(P) + 1/H(Q)
00780  C3 = −C(P)/H(P)−C(Q)/H(Q)
```

```
00790  D = C1*C2-C3*C3
00800  C4 = C2/D
00810  C5 = C1/D
00820  C6 = C3/D
00830  Q1 = C4*A1*A1 + C5*B1*B1 + C6*A1*B1*2
00840  GO = G/E
00850  EO = E
00860  IF A$ = "YES" THEN 910
00870  GO = (E(P)-2)*G(P) + (E(Q)-2)*G(Q)
00880  EO = E(P) + E(Q)-4
00890  GO = GO/EO
00900  PRINT "COMBINED RESIDUAL VARIANCE = "; GO; "ON"; EO;
       "DEGREES OF FREEDOM"
00910  V2 = EO
00920  GOSUB 1080
00930  PRINT "FOR SIGNIFICANCE, F MUST BE GREATER THAN" F
00940  PRINT "F-RATIO ON"; EO; "D.F. IS" Q1/2/GO
00950  PRINT
00960  NEXT J
00970  STOP
00980  AO = 2.30753
00990  A1 = 0.27061
01000  B1 = 0.99229
01010  B2 = 0.04481
01020  T = SQR(-LOG(0.0025))
01030  X1 = T-(AO + A1*T)/(1 + B1*T + B2*T*T)
01040  X2 = (K − 1)*(1-2/9/(K − 1) + X1*SQR(2/9/(K − 1)))^3
01050  PRINT "APPROXIMATE CHI-SQUARE WITH"; K − 1; "D.F. IS" X2;
       "FOR P = 0.05"
01060  PRINT
01070  RETURN
01080  F = V2*((20)^(2/V2)-1)/2
01090  RETURN


01100  END
```

Appendix 6

Listing of program RADIUS (in BASIC) of D. Rodbard for computation of molecular geometric mean radii from molecular weights, assuming sphericity

RADIUS   11:30   22-Sep-78

```
5  PRINT "ENTER MW OF O TO STOP PROGRAM"
6  PRINT
10  REM PROGRAM TO CONVERT MOLECULAR WEIGHT TO R-BAR
15  REM R-BAR = CUBE-ROOT (3/4 V-BAR M / PI N)
20  REM WHERE V-BAR IS PARTIAL SPECIFIC VOLUME
25  REM PI IS 3.14159
26  P = 3.14159
30  PRINT "ENTER V-BAR I. E. PARTIAL SPECIFIC VOLUME
    (USUALLY 0.74)";
40  INPUT V
41  PRINT
42  IF V > 0 THEN 50
44  V = 0.74
50  N = 6.023E23
60  PRINT "ENTER MOLECULAR WEIGHT";
70  INPUT M
71  IF M = 0 THEN 9998
80  R = (3*V*M/4/P/N) ^ (1/3)
82  R = R*1E7
90  PRINT "RADIUS =" R "NANOMETERS"
100  PRINT
110  GOTO 60
9998  STOP
9999  END
```

READY

Appendix 7

Listing of program REFFUB (in BASIC) of D. Rodbard for the calculation of buffer pHs at different values of ionic strength

```
00000  PRINT "ENTER 'STOP' FOR NAME OF ACID TO STOP
       PROGRAM"
00001  PRINT "PROGRAM FOR MONOVALENT WEAK
       ELECTROLYTES"
00002  PRINT "DEVELOPED BY DAVID RODBARD"
00003  PRINT
00004  PRINT "FOR STRONG ACIDS (E.G. HCL) ENTER PK = 0"
00005  PRINT "FOR STRONG BASES (E.G. KOH) ENTER PK = 14"
00006  PRINT "12/18/82: NEW OPTION TO ADJUST PK'S AS
       FUNCTION OF IONIC STRENGTH"
00007  PRINT
00008  READA(1),A(2)
00009  DATA .4884, .5093
00010  REM DAVID RODBARD NICHD RRB NIH BETHESDA MD
       7/22/76
00011  REM I1 IS THE IONIC STRENGTH FOR MEASUREMENT OF PK'S
00012  I1 = .01
00013  PRINT "ASSUME  THAT  PK'S  WERE  MEASURED  AT  IONIC
       STRENGTH"; I1
00014  PRINT
00015  PRINT
00020  REM PROGRAM TO CALCULATE PH AND IONIC STRENGTH FOR
       SPECIFIED BUFFER
00030  PRINT "ENTER NAME AND PK FOR ACID";
00040  INPUT A$,K1
00041  IF A$ = "STOP" THEN 9999
00050  PRINT "ENTER NAME AND PK FOR BASE";
00060  INPUT B$,K2
00070  PRINT "ENTER CONCENTRATIONS OF"; A$; "AND"; B$;
00080  INPUT C1,C2
00090  PRINT "THANK YOU. I WILL NOW CALCULATE PH AND I"
00100  PRINT
00101  IF ABS(K1 − K2) < 20 THEN 110
00105  PRINT "SOMETHING WRONG WITH YOUR PK'S"
00110  X = 10^(K1 − K2)
00120  A = C2/C1*(X − 1)
00121  B = C2/C1 + 1
00122  C = − 1
00123  IF A < > 0 THEN 126
00124  F2 = 1/(C2/C1 + 1)
00125  GOTO 130
00126  F2 = ( − B + SQR(B*B − 4*A*C))/2/A
00130  I = F2*C2
```

```
00131  F1 = F2*C2/C1
00135  IF ABS(F2 − .5) > .4999 THEN 143
00140  P = K2 + LOG((1 − F2)/F2)/LOG(10)
00141  GOTO 150
00143  IF ABS(F1 − .5) < .4999 THEN 147
00144  PRINT "BOTH ACID AND BASE ARE FULLY IONIZED-- NO
       BUFFER CAPACITY"
00145  PRINT "HENCE, PH IS INDETERMINATE"
00146  GOTO 155
00147  P = K1 + LOG(F1/(1 − F1))/LOG(10)
00150  PRINT "THE PH WOULD BE"; P; "AND THE
       IONIC STRENGTH WOULD BE"; I
00155  PRINT "FRACTION IONIZED FOR" A$; "IS"; F1
00157  PRINT "FRACTION IONIZED FOR"; B$; "IS"; F2
00160  PRINT
00170  PRINT
00171  IF C9 = 0 THEN 180
00172  C9 = 0
00173  GOTO 201
00180  PRINT "DO YOU WANT TO CORRECT FOR EFFECT OF IONIC
       STR. ON PK'S < Y/N > ?"
00190  INPUT C$
00200  IF C$ < > "N" THEN 210
00201  PRINT "----------END OF CASE----------"
00202  PRINT
00203  PRINT
00205  GOTO 30
00210  PRINT "ENTER TEMPERATURE";
00211  C9 = 1
00220  INPUT T
00230
00240  J = 1
00250  IF T = 0 THEN 285
00270  IF T < > 25 THEN 900
00280  J = 2
00285  C8 = SQR(I1)/(1 + 1.6*SQR(I1))
00290  K1 = K1 − A(J)*(SQR(I)/(1 + 1.6*SQR(I)) − C8)
00300  K2 = K2 + A(J)*(SQR(I)/(1 + 1.6*SQR(I)) − C8)
00310  PRINT "NEW PK FOR"; A$; "IS"; K1
00320  PRINT "NEW PK FOR"; B$; "IS"; K2; "FOR IONIC
       STRENGTH"; I
00330  PRINT "NOW RECALCULATE PH AND IONIC STRENGTH
       USING NEW PK'S"
00340  GOTO 90
00900  PRINT "SORRY. TEMPERATURE MUST BE 0 OR 25"
00910  GOTO 210
09998  STOP
09999  END
```

Register